Healing with Plants in the

American and Mexican West

Healing with Plants in the American and Mexican West

∎ ∎ ∎

Margarita Artschwager Kay

WITH A FOREWORD BY Andrew Weil

THE UNIVERSITY OF ARIZONA PRESS

TUCSON

Healing with Plants in the American and Mexican West is intended not as medical advice but solely for informational and educational purposes. Do not attempt to use the plants described for medicinal purposes. Please contact a health-care professional about any condition that may require diagnosis or medical attention. The author and publisher disclaim any liability arising directly or indirectly from the use of any of the plants listed in this book.

The University of Arizona Press
© 1996
The Arizona Board of Regents
All Rights Reserved

♾This book is printed on acid-free, archival-quality paper
Manufactured in the United States of America
First printing

Library of Congress Cataloging-in-Publication Data will be found at the end of this book.

British Cataloguing-in-Publication Data
A catalogue record for this book is available from the British Library.

Publication of this book is made possible in part by proceeds of a permanent endowment established with the assistance of a challenge grant from the National Endowment for the Humanities, a federal agency.

To Art, who made all this possible

CONTENTS

ILLUSTRATIONS

FIGURES

MAP

TABLE

FOREWORD

An aloe plant hangs upside down in the door frame of a curio shop in the border town of Nogales, Sonora, placed there to bring prosperity and good fortune. The owner of the shop is convinced that this is a cultural tradition of his people, especially since aloe is one of the commonest cultivars in Mexico, perfectly adapted to the aridity of the Sonoran Desert. But *Aloe* is an African genus, imported by Spanish conquistadores.

When I was studying medicinal plants in the Peruvian Andes, I often watched Indians follow a meal of potatoes—the product of a plant native to Peru—with a chew of coca, another native species. They thought they would get sick if they did not take the two together, because potatoes are 'cold' by nature (regardless of whether they are eaten hot or cold) and coca is 'hot', providing the necessary balance to ward off a 'cold' ailment of the stomach. The classification of foods, herbs, and diseases into categories of 'hot', 'cold', 'moist', and 'dry' is not native to Peru or the New World. It came with the Spanish, who got it from the Moors, since it was an important concept in Arab medicine, which derived it from Galen, who might have taken it from ancient Greece, to which country it might have come from Egypt, where people might have derived it from ancient India.

I have sat with Andean Indians in traditional garb, sharing meals in mountain villages, as they told me about the nature of potatoes and coca, which their people have been cultivating since prehistory: a moving experience. Were it not for ethnomedical scholarship, the methodical inquiry into the origin and development of ideas about health and illness in diverse cultures, one would have no reason to doubt their claims that the 'hot'/'cold' balance of coca and potatoes is a concept as traditional for them as the plants themselves.

When human beings migrate, they take their plants with them—not only familiar foods and beloved ornamentals but also the healing plants that all preindustrial peoples relied on. America has experienced many migrations and is now home to a diversity of ideas and practices about healing. The medicinal plants of many different ethnic groups are available here, in fresh, dried, and extracted forms. Sorting out the cultural origins of these plants is a daunting task for scholars, let alone assessing their uses and possible value in contemporary medicine.

Margarita Kay—anthropologist, nurse, and scholar of ethnomedicine—has taken on just that task with regard to the medicinal plants of the region of North America that she calls the American and Mexican West. As a long-time resident of southern Arizona, I call this region my home as well. As a practitioner of natural medicine who uses many botanical preparations, I have always been interested in Dr. Kay's analyses of the rich traditions of ethnomedicine here.

When I began my studies of ethnobotany in the early 1960s at the Harvard Botanical Museum, we had just come to the end of a cycle of interest in natural products. For most of the next twenty years, it was very difficult to interest foundations, government agencies, private industries, or the general public in medicinal plants, especially those of other cultures. All that has changed in the past few years. Ethnobotany and ethnopharmacology have become fashionable, and consumers are tremendously interested in natural medicine. Health food stores in Tucson are stocked with herbal products, some of which are native to the American and Mexican West. The claims made for these products are sometimes quite at variance with the facts turned up by scholars.

Here is an example: recently a plant native to Amazonia with a wide range of uses by South American Indians has become immensely popular outside of its native region. The plant is *Uncaria tomentosa*, a jungle liana, whose bark is used as an herbal remedy known in Spanish as *uña de gato* and in English as cat's claw. There is little research to back up claims that the plant will heal gastric ulcers, tumors, and arthritis, although some of its constituents appear to enhance aspects of immune function. For several years, pharmacies and street salesmen in Bogotá, Lima, and La Paz have sold as much *uña de gato* as they could get; now sales of the bark and products made from it are booming in the United States, from Miami to New York and Los Angeles.

In western North America the name *cat's claw* is applied to a number of different plants with curved spines. The cat's claw of Margarita Kay's region is *Acacia greggii*, also known as "wait-a-minute" for its notorious habit of holding up hikers with its thorns. Of course, this acacia is unrelated to Amazonian *Uncaria*—and has none of that plant's prominence in native ethnomedicine. Yet along the Mexican-American border, people have been collecting it and other plants known as cat's claw for sale to consumers who have heard about the new miracle herb of that name. At best these products are worthless. Some of them might be toxic. This story shows the danger of relying on common names from one region of the world to another. It also points up the need for good scholarly work in the evolving field of botanical medicine.

Healing with Plants in the American and Mexican West is a welcome addition to the scientific literature in this field. Margarita Kay has tackled an immense amount of historical, cultural, and botanical information and has arranged it into a readable and usable format. I look forward to consulting this book whenever I have questions about the rich ethnobotanical traditions of this distinctive part of the country.

ANDREW WEIL, M.D.

PREFACE

Where can I find out about the herbs my patients are taking? is a question I often hear now that my interest in medicinal plants has become known. Other common questions are Do they really work? Aren't they very dangerous? Are they "just" placebo? Do Mexicans and Indians use different remedies? Why do they have different names? What are the "correct" names for these herbs? It is hard to find answers to such questions about natural remedies used by Mexican Americans from southern Arizona and New Mexico and by peoples of the Mexican states of Sonora, Chihuahua, Baja California Norte, and Baja California Sur—primarily desert regions.

I have recommended Karen Cowan Ford's excellent *Las Yerbas de la Gente: A Study of Hispano-American Medicinal Plants* (1975), but it does not extend coverage to the Mexican regions. Daniel Moerman's two-volume index, *Medicinal Plants of Native America* (1986), outlines the plant use of the Navajo, Apache, Hopi, Zuni, and Pima who live in Arizona and New Mexico but not that of the (sometimes related) Native Americans of the Mexican states. Michael Moore in *Medicinal Plants of the Mountain West* (1979) and *Medicinal Plants of the Desert and Canyon West* (1989) discusses many of the same plants that I deal with, but these books are directed to different readers: those who would actually collect, prepare, and use the plants for treating themselves. Otherwise, data on medicinal plants in the American and Mexican West are mostly scattered, hidden in erudite journals or ethnographies.

In this book I have attempted to fill the gap by collating information from the many ethnographies and articles with data from hundreds of interviews that I conducted with people who use or dispense remedies made from plants. As a nurse, I want to know how people treat themselves when they are sick. As an anthropologist, I am interested in what people have used through time, universal materials with which to treat illness. I hope to contribute to the theory of medical anthropology by exploring the evolution of the Mexican American domestic pharmacopoeia: how it has developed from available plant materials, how culture contact has affected it, which plant uses became obsolete and which persisted after competing biomedicines arrived, whether a shared geographical biome or related language use might explain similarities.

My interest in medicinal plants was first awakened when I began work on my dissertation, *Health and Illness in the Barrio: Women's Point of View,* for the Mexican American women who were my informants used *yerbas,* or plant remedies. More orientation to ethnobotany came when I took the parts of medicinal plants to botanists at the University of Arizona Herbarium, under the direction of Charles T. Mason, for comparison with voucher specimens. I also learned from Robert Bye and Edelmira Linares, Gary Paul Nabhan, Willard Van Asdall, and Joseph Laferrière. Through Andrew Weil I was introduced to members of the American Botanical Council and the Herb Research Foundation and their publication, *HerbalGram.*

I have worked at various times in the Wellcome Institute of the History of Medicine, London, where I read incunabula and facsimiles of early works from Europe and the New World. I am grateful for the help of everyone there and give special thanks to Robin Price, deputy librarian of the Americanist collection. I conducted historical research also at the Huntington Library in Pasadena, California; the National Library of Medicine in Bethesda, Maryland; the Library of Congress in Washington, D.C.; and the University of Arizona's Main Library, Special Collections Library, Arizona Health Sciences Center Library, and Anthropology Library in Tucson. The many librarians, beginning with Marilynn Smith (now retired from the Arizona Health Sciences Center Library) were always helpful.

In addition, I should like to thank many others for their support in this endeavor. Funds to assist in the research were given by the University of Arizona's Mexican American Studies and Research Center, Columbus Quincentennial Program, Southwest Center, and College of Nursing, which granted me two sabbatical leaves for this study. Phytochemical information came from NAPRALERT, a database of natural products maintained by the Program for Collaborative Research in the Pharmaceutical Sciences within the Department of Medicinal Chemistry and Pharmacognosy in the College of Pharmacy of the University of Illinois at Chicago. I have also used texts on pharmacognosy as well as the *Journal of Ethnopharmacology.*

Others whose help I must acknowledge are my former teachers at the University of Arizona, including James Officer and the late Edward H. Spicer and Thomas B. Hinton, all ethnohistorians; Mary O'Connor, who oriented me to eighteenth-century calligraphy and to contemporary Mayo; Charles Polzer, S.J., who gave me letters of introduction to European libraries and historians and to Edward Burrus, S.J., late of the Vatican library. I have seen which plants are used for medicine through the help of many individuals (doctors, nurses, *curanderos, abuelas, viejas*), including Mary Helen Allison, Georgiana Boyer, Esther Contreras, Eva Magallanes de Contreras, Gloria Giffords, Josefina Lizárraga, Sandra Gonzalez Marshall, Ray Martínez, Agustus Ortíz, Margarita Redondo, Anita Stafford, Emilio Verdugo, Carmen Altamirano Wilson, Michael Winkelman, and many others. Often people have advised me, "You should talk to so-and-so about the best *yerbas medicinales*," and so I have done. A network of colleagues in the field of medical anthropology is developing the theory base; I should list fifteen hundred others but will mention only Carmen Anzures y Bolaños, George Foster, Jody Glittenberg, Daniel Moerman, Mary O'Connor, Bernard Ortiz de Montellano, and Lola Romanucci-Ross.

Joseph Laferrière wrote most of the plant descriptions and reviewed the plant taxonomy for this book. Cornelius Steelink reviewed the section on phytochemistry in chapter 2. Gary Kabakoff wrote the programs that enabled me to organize the information in Data Perfect and organized the manuscript through several rewritings. I also thank the anonymous reviewers; their suggestions directed the changes. I thank all of the people at the University of Arizona Press, especially Christine Szuter and Sally Bennett.

Of course, it all started long ago. I wish I could thank my father, Ernst Artschwager, who was a plant anatomist and taxonomist, and my mother, Eugenia Artschwager, who was a linguist and artist. She started me in my love of words, and my brother, Richard Artschwager, in his love of art: I can thank him for this book's cover illustration. Most of all, thank you Art Kay, for letting me drag you through herbal gardens here and in Mexico, Europe, and Africa and for teaching me how to write about what I saw.

I hope that *Healing with Plants in the American and Mexican West* can be a useful source of information for doctors, nurses, nutritionists, pharmacologists, toxicologists, and other health-care providers. My purpose is to present facts with neither advocacy nor wholesale condemnation about healing traditions that use plant remedies. Further, I hope that my work will encourage investigation of the biological activity and chemical constituents of other plants of the American and Mexican West.

Healing with Plants in the

American and Mexican West

Introduction

Guadalupe Fraser noticed another infection on her foot of the kind that has plagued her since she developed diabetes. To treat it, she took a leaf of a *sábila* (*Aloe*, aloe vera) plant from her yard, removed the skin, cut it in half vertically, and heated it briefly over the kitchen stove. Then she applied the oozing mass directly to the infected area. Next she took a leaf of *maguey* (*Agave*, century plant), which she chewed, then discarded the skin and applied the pulp to the same infected area. These materials were held in place with a bandage. She repeated this process three times daily. That will cure it, she said to herself.

Guadalupe Fraser is a Mexican American woman in Tucson, Arizona, whose use of Old and New World medicinal plants reflects her heritage—Spanish, Indian, and Anglo. Her doctor might be horrified to learn that she is treating herself in this way, but Guadalupe says it works.

The Columbian encounter—far away in the Caribbean and five hundred years earlier—ushered in this scenario in the far northwestern regions of today's Mexico and the southwestern United States. Guadalupe selected both *maguey*, an ancient New World remedy for wounds, and aloe, brought from the Old World. This choice was a synthesis of medicinal plant knowledge reflecting the new pharmacopoeia of healing created after the Conquest.

To emphasize the common cultural use of plants as medicines by peoples of the southwestern United States and northwestern Mexico, I call the area the American and Mexican West, a region named at different times the Chichimeca, Northwest New Spain, Arid North America, the Greater Southwest, the Desert Borderlands, and México Norteño. The area now includes much of the U.S. states of Arizona, New Mexico, and southern California and the Mexican states of Chihuahua, Sonora, Baja California Norte, and Baja California Sur.

The plant pharmacopoeia used to this day—from what we know of the earliest documented medicinal plant use—includes herbs of Renaissance Europeans; remedies of sixteenth-century Aztecs, who are said to have come from the north and west of Mexico; and plant medicines of eighteenth-century peoples of northwestern New Spain. These herbs are still employed by the modern-day descendants of the earlier peoples of the area: upper Pima, including Tohono O'odham; lower Pima, including Ópata and their descendants; Tepehuan; Mountain Pima; Yaqui; Mayo; Tarahumara; Warijio; Seri; Paipai; other Baja Californians; *norteños,* that is, Mexicans from the Northwest; and Mexican Americans.

This book generally follows common usage of the word *herb* in referring to plant parts from any life form that are used as medicines. These may be bark, root, stem, flowers, fruit, seed, or sap. In the botanical descriptions of part 2, the word *herbaceous* is used to describe a plant whose stem does not produce woody, persistent tissue and generally dies back at the end of each growing season.

I have looked for answers to questions such as What are the plant pharmacopoeias of the American and Mexican West? Have different ethnic groups used the same species or other species of the same genus? Does common language, geographical contiguity, or shared habitat best explain shared use of medicinal plants? What is the history of use of specific plants? What do the names of plants tell about their histories? How has medicinal use of plants in specific genera resembled and differed from use of related plants in other parts of the world? What do the uses of plants reveal about the illness beliefs of the people who employ them?

The archaeology of medicinal plants shows that the plants used for treating illnesses today have not been haphazardly chosen but instead exhibit long histories. Each cultural group depends upon the plant materials that are available, designating specific plants as useful medicine—yet only a tiny fraction of the 250,000 to 750,000 plants estimated to exist are used as medicines in the pharmacopeias of various societies. Further, often the same genera of plants are represented in the medicines of peoples living in widely separated parts of the world. Why the same or closely related plants—only a tiny fraction of the possible number—have been selected everywhere is an intriguing question.

Using the ethnohistorical method,[1] one may apply our present behavior to what might have occurred in earliest times. People must have exchanged knowledge of healing with plants even before they began to keep records of such events. Because, for example, travelers today take medicines with them when they go to unfamiliar places so that they are not caught without, does it not seem likely that the earliest people coming to the American and Mexican West would have provided themselves with familiar remedies? Or once settling in a new area, might they have recognized plants whose healing properties they knew? Other healing plants would be newly discovered, newly tried, perhaps converging with developments elsewhere. This knowledge would then be passed on to recent arrivals or other peoples met in the area. Indeed, my research shows that within the Mexican and American West wherever the same plants grow (native or naturalized), their medicinal uses have been exploited by various peoples.

People deal with events according to their cultures. Nevertheless, although they give different labels to and explanations for health problems, they use a common core of plant, animal, and mineral materials for treatment. For the anthropologist, the history of use of a specific plant can thus lead to understanding what is universal in the ways of healing throughout sundry cultures. People also learn to perceive, interpret, and respond to illnesses in various ways, often determined by their culture. For example, except persons with rare neurological problems all people perceive pain. However, their interpretations of pain differ: the pain may mean supernatural transgression, environmental variation, natural phenomena, or disease.

Health-care providers need to appreciate how culture affects what their patients do about health-care problems. They should become acquainted with the alternative therapies that are sought in this region, sometimes clandestinely (and often in conjunction with biomedically prescribed therapy),[2] and should have facts that can help them decide whether to incorporate such therapies or advise against them. Acknowledgment of other kinds of health care should not mean romanticizing unsafe practices. Providers of biomedicine who come to respect or at least accept the fact of ethnomedical (folk) knowledge and practice rather than condemning self-care wholesale will provide safer care to their patients. This book is addressed particularly to those who see themselves as medical culture brokers, facilitating and clarifying communication between the various professionals—physicians, pharmacists, therapists, and especially nurses—and their patients.

The need for this book became apparent to me after a colleague, Mary Helen Allison, and I conducted a workshop on commonly used medicinal plants for the National Hispanic Council on Aging. A nurse practitioner who had attended this workshop told me some time afterward that she was able to apply what she had learned from us to the care of a diabetic patient whose blood glucose level was fluctuating dangerously. When feeling weak, this patient was taking the root of *matarique* (*Psacalium decompositum*, Indian-plantain) in addition to the medically prescribed hypoglycemic (Tolbutamide), in effect giving himself a double dose, for *matarique* also has hypoglycemic activity. Also, the nurse practitioner learned from our workshop that the root poses another danger: it contains pyrrolizidine alkaloids, chemical compounds that when ingested over a period of time can cause severe damage to the liver and lungs. Nor is this plant the only one he might have taken, since more than twenty plants are commonly used to treat diabetes, a common problem for native populations, in traditional medicines of the American and Mexican West. In addition to the safety concerns, the nurse practitioner learned the skill of treatment negotiation so that she could provide an atmosphere in which the patient felt comfortable in revealing his use of the medicinal herb, and she realized how this might be accommodated in his therapy.

Who uses plants today for healing? Some people employ medicinal plants when their ailment seems not to be serious enough to warrant a physician's examination. Mexicans and Mexican Americans have a long tradition of healing by housewives, passed orally from their grandmothers for what they consider *enfermedades benignas,* minor ills. Other people use traditional medicine to avoid expensive biomedical care, learning from *yerberos* (herbal specialists), home doctoring books, or clerks in *botanicas* (herb stores). Undocumented workers, afraid of encounters with immigration officials, bring *remedios* (folk remedies) across the border or recognize them in the agricultural fields where they labor. Cost is a factor for many people throughout the world. Locally growing plants may be free; packaged herbs are usually a great deal less expensive than pharmaceutical medicines. Other people will go to great efforts to obtain traditional remedies, some because they distrust "chemical" medicine—believing natural compounds to be superior to synthesized ones—and others who are unable to take certain medicines because of sensitivity to the chemicals. Persons with chronic or terminal disease will try remedies they formerly scorned. Medicinal plants are an integral part of *curanderismo,* traditional healing. They are also used in teas and baths for treatments in *espiritismo,* spiritism (Finkler 1983), a movement that is attracting increasing numbers of people, especially Mexicans and Mexican Americans. Some biomedical physicians prefer treating with natural products (Weil 1990, 1995). Homeopathic physicians prescribe many medicines derived from plants, following a different theory of therapeutics.[3] Naturopathic doctors, many of whom are highly knowledgeable in biochemistry, use plant preparations for therapy (Murray and Pizzorno 1991). All these reasons add up to a significant interest in medicinal plants.

Recognizing that history, culture, and ecology all influence plant use, I propose that the pharmacopoeias of the American and Mexican West developed in the following way. First, the ecology of the region determined which plants could be found. The medicinal effects of plants growing in areas such as the unique environments of the Sonoran or Chihuahuan Desert were discovered, then passed on to other groups outside the area, perhaps to those related by language. In this way plants perceived to have more effective medicinal actions or other attractive features were added to the repertoire. When one group came into contact with another, each learned new information about plants as medicine—as when Europeans brought their medical culture into contact with indigenous societies. European colonists accepted some indigenous remedies to replace the ones that they had always used but could no longer obtain. Conversely, native peoples adopted plants used in the European official medicine of the time, in part because the introducers of those remedies—Spanish conquistadores, priests, and doctors—were perceived as having great power (power that extended to their medicines) or because some gentler-acting medicinal plants from the Old World could replace desert-adapted growth. Isolated societies kept their own plant medicines, while also incorporating those new ones that were available, perhaps through the traditional networks of trade. Change was also forced, for the powerful conquerors reviled the old as diabolical and considered the specialists in native medicine to be agents of the Devil. However, old ideas did not necessarily disappear when these traditional healers were repressed.[4]

Likewise, when individuals choose medicine today, they may continue to use traditional treatments but also may be attracted by new remedies because of convenience, effectiveness, safety, or more philosophical concerns. A plant medicine may be superseded by another that is easier to obtain, making it no longer necessary to search out a cactus for its root when one can get a weed from the backyard. A less-effective plant medicine for headache may be replaced when a better one is neatly packaged and readily available at the grocery store. A dangerous plant purgative may be dropped when a mild cathartic is procurable. Old medicines may be discarded, replaced by the new; or they may go underground, continuing in a healing tradition parallel to the official system but with a different theory of disease, different kinds of persons designated to conduct treatments, and different materials for healing (see Kay 1978). All of these processes have taken place in the American and Mexican West.

This model developed from information in two unpublished data assemblages that I have compiled through the years: Ethnoagents in Women's Health Care (see Kay and Yoder 1987) and Medicinal Plants of the American and Mexican West. Additional data have been abstracted from ethnographies, ethnobotanies, articles, and interviews.[5] The records include the most commonly reported plants of the American and Mexican West, both historically and currently. I used the stock inventories of medicinal herb stores and grocery stores in Tucson in October 1995 to verify the present availability of plants. The plant parts are packaged in packets of one to two ounces and are labeled with the common Spanish name (occasionally the botanical name is listed on order forms; however, these names may be misspelled, archaic, or incorrect).

In part 1 of this book I present ethnomedical information including the historical uses of healing plants and an overview of the ethnic groups of the region (chapter 1); a summary of the naming, selection, and perceived actions of medicinal plants (chapter 2); and an assessment of the conditions treated by plants, both those which are recognized in biomedicine, such as arthritis, and folk illnesses such as *pasmo* (chapter 3), as well as the specialized illnesses of women and children (chapter 4). The one hundred most commonly reported plants of the American and Mexican West are presented in detail in part 2.

Information about the action of a plant includes folk efficacy (what people believe that using the plant accomplishes) and folk properties (how the plant is believed to heal), which can be crucial in explaining why these plants continue to be used in healing, as well as the probable explanation of efficacy in biomedical terms, the pharmacology of its constituents.

For more than a century researchers have been studying natural products; about 121 clinically useful prescription drugs have been derived from higher plants.[6] Although the relatively recent science of chemical synthesis eclipsed the use of plants as pharmaceuticals, the twentieth century has seen increased interest in the study of medical ethnobotany. Plant scientists published many ethnobotanies beginning in the 1930s and 1940s. Anthropologists like myself are other scientists concerned with medical ethnobotany; the study of plant use can lead to understanding culture contact, change, economics, religion, politics, and ethnic identity.

The medical anthropology of Mexican Americans and other Hispanics has been studied by too many scholars to list here,[7] but few of these analyses of traditional medicine by anthropologists have reached biomedical practitioners; even fewer give full descriptions of how plants are used as remedies. Moreover, much of the ethnobotanical study has been directed toward plants of the northeastern United States.[8] Thus, little information is available on the native plants of the desert that are used for medicine in the American and Mexican West.

Many medical ethnobotanical reports state that the author performed the work to prevent this information from disappearing entirely. In fact the knowledge never disappeared in this region, although the popularity of plant medicine waxes and wanes. Today there is new interest—even a sense of crisis—as whole biomes (major biological communities) are being destroyed. Searches of the Amazonian rain forest for plants that could provide new drugs are appropriately being recommended (Balick 1990), but we also need to learn about the potential of plants in our backyard that have already been used in folk tradition. Plants that have been used for hundreds of years in the American and Mexican West remain virtually ignored even though various studies suggest that when plants are screened not randomly but on the basis of their ethnobotanical use between 20 and 52 percent show biological activity (Huxtable 1992). Because new treatments and cures are needed, plants of the desert should be brought to the attention of pharmaceutical scientists.

Today many health-care providers deride using plants for medicine. However, in the United States, plant medicine is a billion-dollar business. Health food stores, *yerberias,* and large grocery chains carry hundreds of different medicinal herbs. Mexican Americans can purchase packets of flowers, roots, or leaves labeled with Spanish common names, together with traditional booklets in Spanish describing purpose and preparation. The inventory is not static: for

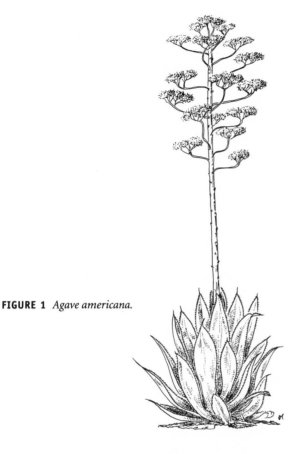

FIGURE 1 *Agave americana.*

example, the plant *guareke* (*Ibervillea sonorae,* coyote melon), reported by Father Ignatz Pfefferkorn in the mid-eighteenth century as good for healing wounds, is used again but now is recommended to treat cholesterol and diabetes, twentieth-century concerns.

Mexico has had a sustained interest in studying its traditional medicine. Indeed, IMEPLAN, the Mexican Institute for the Study of Medicinal Plants, and COPLAMAR, the national coordinated plan for depressed regions and marginal groups, combined under the Mexican Social Security Institute (IMSS) for joint studies. Their publications provide valuable data for contemporary and comparative use of medicinal plants in each state of Mexico, including those of the American and Mexican West.[9] However, despite the official interest by IMEPLAN, IMSS, COPLAMAR, and other institutions, the average Mexican physician is likely to have little knowledge of folk medicine (see Anzures y Bolaños 1983). The traditional medicine of northwestern Mexico (which is "so

far from God, so close to the United States," as the saying goes) may indeed be
the least known.

The health-care provider has very practical interests in these plants. Do
medicinal plants effectively treat illness? The answer depends in part on defin-
itions of the terms *illness* and *effective*. A treatment may be considered effective
if it results in the diminution of symptoms, resolution of discomfort, or resto-
ration of health (Etkin 1988:25)—or if it acts on the body as expected by the
people using it. This is *emic* efficacy, that is, it meets the culturally defined expectations of the healer, patient, and social group, confirming and reaffirming
shared beliefs about the nature of health: for example, if it causes the diaphore-
sis (sweating), emesis (vomiting), diuresis (urination), or purgation (evacuation of the bowels) believed to eliminate the illness (Etkin 1986). Therapy for
conditions categorized in the folk nosology as 'cold' diseases may include treatment by plants once codified as 'hot' even though the classification system may
have been forgotten (Kay and Yoder 1987), a practice known as archaic theoret-
ical efficacy. In addition, some of the plants used in the American and Mexican
West owe some of their presumed efficacy to magical authority in addition to
physical power: for example, *albahaca* (*Ocimum basilicum,* basil) counteracts
witches as well as treats earache and women's problems, and *pirul* (*Schinus
molle,* pepper tree) helps backache and also protects the home. Some researchers have argued that folk efficacy should be evaluated by the degree to which
the plants produce the wanted effects within the user's own system of medical knowledge (Ortiz de Montellano and Browner 1985) and that a plant can
be scored as to both its emic (cultural) and its etic (biomedical) effectiveness
(Browner, Ortiz de Montellano, and Rubel 1988).

To suggest the effectiveness of the included medicinal plants in biomedical
standards, I am relying on recent pharmacological studies that have been conducted for the reverse of reasons why plants were first so studied. For more
than 100 years, plants have been analyzed to learn the one active ingredient in
each plant demonstrated to be responsible for its actions on human physiology
so that this compound may be synthesized. In 1976, the World Health Organization was requested to take on the subject of traditional medicine in support
of the Declaration of Alma-Ata: "Health for all in the year 2000."[10] Subsequently, multinational activity has been directed toward learning the healing
properties of plants. The phytochemical information in part 2 I synthesized
from NAPRALERT, a data base of natural products literature at the University
of Illinois (see bibliographic essay) that offers biological and chemical profiles
abstracted from approximately 85,000 journals and other reference works
(Farnsworth 1983). However, when each plant has a list of "mind-numbing
length" (Brown 1987:7) of chemicals, it is difficult to be certain which compounds might be responsible for the effect. The action of even known toxins
may be attenuated, made more potent, or palliated by other bioactive con-

stituents in the same plant or a mixture of plants, or by some form of preparation of the medicine, such as cooking.[11]

For these kinds of reasons, few Western biomedical personnel have encouraged self-care with plants. Biomedical health-care providers are uneasy with
plants as medicines because each has hundreds of compounds: all, some, or
only one may contribute to observable effects.

For safety, one generally cannot recommend self-medication by those who
do not have a great deal of knowledge about medicinal plants, and too little information is available for most laypeople. Plants containing compounds that
are strong enough to effect pathophysiological processes in humans are by definition toxic. Moreover, the efficacy of the plant in obtaining a cure is rarely
evaluated through the randomized control research utilized in biomedical
studies (Anderson 1991). Complete biomedical investigation of a plant remedy
that will lead to its approval and registration of its active ingredient by the
United States Food and Drug Administration (FDA) is such a lengthy and elab-
orate process that it is rarely undertaken. Indeed, it was estimated in 1993 that
to get a new drug approved cost $231,000,000 (Blumenthal 1993a), although I
am told that this frequently quoted estimate is inflated by including the costs
of investigating all plants submitted for possible study. According to phyto-
chemist Barbara Timmerman, the cost of analysis of one plant is closer to
$100,000. The FDA has approved some medicinal plants as foods "Generally
Recognized As Safe" but at this time does not permit the labeling of any for
medicinal uses. Nevertheless, the Office of Alternative Medicine was established in the National Institutes of Health in 1993; part of its mission is to explore the appropriate regulation of botanicals as medicine. Next, authorized by
the Dietary Supplementary Health and Education Act of 1994, a commission
was appointed in 1995 to study claims of medicinal use on labels and to make
recommendations for approving them.

At this writing, using plants for medicines has been revitalized—seen as
trendy "alternative medicine" by some, as hope for the desperate, economic
necessity, or cultural comfort by others. It may be hoped, then, that this book
will draw the attention of health-care providers and others interested in better
understanding this increasingly popular yet potentially dangerous practice.
Two caveats are essential here: first, the information is insufficient to allow
the health-care provider to prescribe any of the plant medicines as treatment,
and second, in no way is the material suitable for the layperson to initiate self-
treatment.

1.

1 ▪ Ethnohistory

The people of the American and Mexican West still employ ancient plant remedies that they believe are useful, easily available, and not superseded by new medicines. Since prehistoric times the same kinds of plants have been used in various places in the world for healing, the particular species limited only by their physical requirements. The present plant pharmacopoeias of the American and Mexican West, while constrained by desert ecology, reflect numerous historical and cultural influences; indeed, they have evolved through cultural exchange since before the time of Columbus in Europe and in the New World, a process continued in colonial Mexico and into the twentieth century. Perhaps the most direct means to understand the use of medicinal plants is to ask the peoples of the region or to learn from the scientists who have studied and continue to study their medical ethnobotanies.

Healing with Plants in Prehistory

How did mankind ever find out that certain plants can heal? Theories abound: prehistoric humans learned by serendipity and then diffused this knowledge throughout widespread networks of trade or language groups; therapeutic properties of certain plants were independently discovered throughout the world; humans acquired knowledge of medicinal plants by observing animals that seemed to use plants when they were sick; humans starving and forced to eat certain plants might have discovered their healing properties.[1] All these processes must have contributed to the development of plant pharmacopoeias.

Let us start then with mankind's beginnings here. Thirty thousand and more years ago when climate change opened corridors from the Asian continent to North America, people began to come from Siberia to the New World. Using linguistic, dental, and genetic evidence, researchers have estimated that these early immigrants came in three distinct waves. According to this theory, speakers of Amerind languages came in the first wave, eventually encompassing those who spoke Uto-Aztecan languages—whose descendants included Pima, Tohono O'odham, and Pueblo peoples in the northern limits and Aztecs in the southern limits—and Hokan languages, including Paipai and Seri. Next came the speakers of Na-Déné languages, Athabascans represented by the Navajo and Apache in our area. The last were Eskimo-Aleuts.[2]

These waves of prehistoric peoples likely brought either medicines made from plants or information about which plants were medicines, recognizing those that they had used for healing before coming to the new lands.[3] In the millennia that followed, as immigrants pushed southward into new lands, they would have found new plants that could be used as medicines. Doubtless many plants were discovered independently in different places.

There is a long tradition of belief that animals first discovered the medicinal values of plants. The medical "discoveries" of swallows, dogs, and deer, for example, were described in the first century A.D.[4] One example of animal medicinal use is *Ligusticum* (lovage), a plant still employed in the American and Mexican West. According to legend, the Navajo learned from bears that a *Ligusticum* species cures worms, infections, and stomach ills. Alaska's Kodiak bears dig up, chew, swallow, or rub their fur with the roots, and brown bears at the Colorado Springs Zoo relish *Ligusticum* and also apply the chewed root to their face and fur (Begley 1992:53–54). Perhaps Asiatic bears liked *Ligusticum*, too, and taught the proto Na-Déné, but in any case this example lends support to the idea that the Athabascans brought medicinal plant beliefs with them when they traveled from the far north to the Southwest.

As another alternative, might starving people have discovered the medicinal properties of certain plants when forced to eat something new? Many plants now listed as famine foods and identified from coprolites, ancient human feces, are also known as medicinals (Minnis 1991). Indeed, the idea that a plant can provide *either* food or medicine, but not both, is recent. Early immigrants must have tried to use whatever was available for both food and medicine when familiar plants were absent.

Healing with plants is indeed ancient, but archaeological evidence about the origins of medicinal plant use is speculative. The earliest indication of human use of medicinal plants comes from fossil pollens found on skeletal remains of Neanderthals from a site in northern Iraq (Shanidar), carbon-dated at more than 60,000 years ago. The largest quantities of pollen came from seven kinds of plants that are known throughout much of the world today for their medicinal properties,[5] including five that are employed today as medicines in the American and Mexican West: *Achillea, Centaurea, Senecio, Ephedra,* and *Althaea.* The use of *Achillea* (*plumajillo*, yarrow) species can be traced from Siberians and Aleutians, Inuits, and Tlingit Athabascans of Alaska to Mexican Americans of Colorado, who employ the plant as did the ancient Aztecs—for wounds as well as stomachache, cough, and urinary complaints. Rarer in the American and Mexican West is the use of flowers of a *Centaurea* (*cardo santo,* star thistle), which are chewed by Spanish American women of New Mexico during childbirth labor, while the Yaqui use the related thistle *Cnicus,* which they also call *cardo santo. Senecio* (*calancapatli,* groundsel) is used by many peoples here for sores and is also used by Aleuts. *Ephedra* (*cañutillo del campo,* joint fir) is used for urinary complaints by Paipai, Mexicans, and Mexican Americans. Plant relatives of *Althaea* in the Malvaceae—in the American and Mexican West represented by *Malva* (mallow)—are known by many ethnic groups as having a soothing and cooling action (Lietava 1992:265).

A site in Chile dated 13,000 years ago suggests a long history of healing with plants indigenous to the New World. The remains of sixteen plants were found preserved in the peat bog of this early human site: Tom Dillehay, the archaeologist, inferred medicinal use of these plants because only the medicinal parts of each plant, such as roots or leaves, were present. Mapuche Indians still gather many of these plants today as then, from far distances, and use them to treat cold ailments or pulmonary diseases. For example, *Pneumus boldo* (*boldo*) leaves are used in an herbal tea to treat common colds, congestion, and stomach disorders (Dillehay 1987). This plant is employed widely today by Mexican Americans in the American and Mexican West. *Lycopodium* (*licopodio,* club moss)—used by people as widely distributed in time and space as ancient Celtic druids and present-day Alaskans—was also well represented in the samples tested. Other common species included plants in the *Atriplex, Centaurium, Equisetum,* and *Mimulus* genera, all used medicinally in the American and Mexican West. Of sixty-eight plant species recovered from the 13,000-year-old site, thirty-two still have medicinal uses, according to the project botanist, Carlos Ramírez (1989).

In the American and Mexican West, other paleoarchaeological studies of pollen likewise suggest that plants were used for healing. Pollens have been found in fossilized feces, or coprolites, in southwestern cave sites dating from A.D. 200 to 1000. Plants used as medicine today whose pollen has been found in large amounts in these prehistoric sites include *Ephedra* (*cañutillo,* horsetail), *Larrea* (*gobernadora,* creosote bush), and *Salix* (*sauce,* willow). *Chenopodium* (*epazote,* goosefoot) was also found in another study.[6] However, no southwestern ethnographic source identifies the flowers of *Salix,* the source of pollens, as having been used for medicinal purposes;[7] nor for that matter are the flowers of *Ephedra, Chenopodium,* or *Larrea* the medicinal parts.

Pre-Columbian Exchange in the American and Mexican West

Through trade, the peoples of ancient Mesoamerica and the American and Mexican West might have exchanged knowledge about medicinal plants.[8]

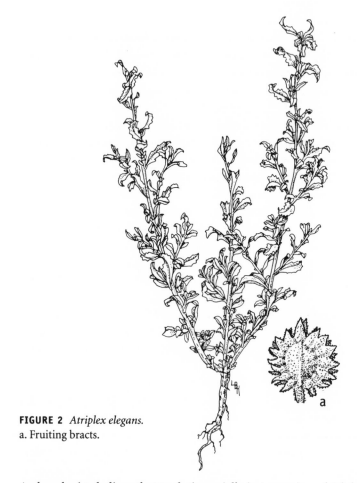

FIGURE 2 *Atriplex elegans.*
a. Fruiting bracts.

Archaeologists believe that trade (especially in turquoise, which had great religious significance in Mesoamerica) helped to create extensive commercial and cultural exchanges between Mesoamerica and the American Southwest long before the Spanish conquistadores arrived (Harbottle and Weigand 1992:78): it seems reasonable that curing materials were interchanged in trade. Shortly before the conquistadores arrived, Athabascan peoples (the Navajo and Apache) came to the Southwest. They too may have recognized plants similar to those they used as medicines in the far north, and they may also have learned which plants were medicinals from trade and interaction, for today they use many of the same plants as their Mexican American neighbors.

Why would different peoples throughout the world use the same plants, or plants that are closely related by genus or by family? A study of the world geography of plants shows that related species of plants often are indigenous to the same kinds of environments. Thus, the arid regions of the American and Mexican West share many medicinal plant genera with other desert lands in places such as Afghanistan, Ethiopia, India, Israel, Peru, Sudan, Syria (Ayensu 1979). Siberia and Alaska, although cold, are nevertheless arid regions and also share genera with the American and Mexican West.[9] Yet although plants in the same genus or related genera of plants may be used for medicine in many areas of the world, the particular species in the genera are likely to differ. For example, plants in the *Acacia* genus are used medicinally by many peoples, but each environment favors a different species (see part 2). The common use of plant genera and species by people in widely separated regions is also at times due to human use rather than environmental limitations: some species are purposefully introduced because of their medicinal value, while other species invade, to become widely dispersed (Good 1974:187)—many of these "weeds" are used as medicines. Their success at invading may be because of pharmacologically active chemicals that give them a biological advantage over their competitors, making them unpalatable to predators or inhibiting the growth of other plants.

The Plant Pharmacy of Europe

The plant pharmacopoeia of the American and Mexican West derives from two traditions, indigenous American and European. At the time of the Columbian encounter, medical documents used in Spain typified European knowledge, little changed since antiquity. The earliest record of medicinal plant use that we have today is an Egyptian papyrus written in the sixteenth century B.C. It included *Sambucus*, elderberry, which is one of the most popular medicinals in the American and Mexican West. Ancient compilers of herbal knowledge included the Greeks Hippocrates, Aristotle, Theophrastus, Dioscorides, and Galen and the Romans Pliny and Celsus in the early centuries of the Christian era. Their books, ancient summaries of even older writings, remained the mainstay of medical theory, passing into Arabic hands during medieval centuries. Arabic interpreters of these Greek and Roman writers included Avicenna, the Persian Rhazes, and the Jewish Maimonides, whose works were valued in Moorish Spain. With the onset of the Crusades, Europe was again exposed to ancient medical theory.

Representative of Spanish medical writings of the early fifteenth century is *Menor daño de la medicina de Alonso de Chirino*, "Alonso de Chirino's Least Damage by Medicine" (Herrera 1973), the title recalling the Hippocratean aphorism "First, do no harm." The book gives tables of rules for a healthy regimen, followed by treatments of illnesses, like the medieval *Tacuinum Sanitatis* (Arano 1976), that were translated from Latin versions of Arabic works. Late in the fifteenth century, books such as *El sumario de la medicina con un tratado de las pestiferas bubas*, "Summary of Medicine with a Treatise on Bubonic Plague," by Francisco López de Villalobos ([1499] 1973) appeared. Both Alonso de Chirino and López de Villalobos recommend the same remedies that had appeared in Dioscorides' *Herbal* and Pliny's *Natural History* during the first century A.D.

The theory of medicine was based on a system of classification that assigned properties of hot/cold/moist/dry to persons, their four "humors" (blood, hot and moist; phlegm, cold and moist; yellow bile, hot and dry; and black bile, cold and dry), illnesses, and remedies. Although this theory of disease has been completely discredited by scientific study, some observations might explain how the theory was devised (MacFarlane 1962). The Greeks noted that the blood separated into layers when obtained from sick patients, the four humors being represented by these separated parts of the blood. According to modern examination of this theory, the sedimented red cells can be divided into a dark or "melancholic" humor and a red or "sanguine" humor, depending on the degree of oxygenation, while the upper cell-free fibrin clot would constitute the phlegmatic humor, and the supernatant serum, the choleric humor. Humoral theory remains strong in folk belief throughout the world.[10]

Theorists during the European Renaissance varied in their interpretations of humoral theory—and in nomenclature—as each attempted to provide a translation that was most faithful to the ancients (Siraisi 1985, 1990). Many herbals were printed in vernacular languages throughout Europe, repeating much of the same information from Theophrastus, Dioscorides, Pliny, and Galen. In Germany, writers included the emulated Hieronymus Bock; in the Netherlands, Rembert Dodoens and Charles de L'Écluse. In England, the names William Turner and John Gerard are foremost. Religious and political competition between scholars such as John Goodyer (who translated Dioscorides) and Turner (called the "father of English botany" because he wrote the earliest English herbals, *Libellus de Re Herbaria Novus* in 1538 and *The Names of Herbes* in 1548) stirred the intellectual pot (Jackson 1965).

While these books represented scientific knowledge of the time, a competing folk tradition was also active. Sir Robert Burton ([1620] 1927:563) aptly wrote, "Many an old wife or country woman doth often more good with a few known and common garden herbs than our bombast Physicians with all their prodigious, sumptuous, farfetched, rare, conjectural medicines." Pliny had said much the same about women herbalists in the first century A.D.[11] Through the centuries, these "known and common garden herbs" have been used in systems of medicine as alternatives to the prevailing official medicine.

The Plant Pharmacy of the New World

Early Studies of New World Plants

The Renaissance was not only a revival of ancient knowledge, it was an eagerness for discovering new things: Columbus's travels to the New World occurred when Europeans were greatly interested in learning new ways in healing. The study of medicine and the study of botany were virtually the same topic in Europe's age of exploration. Europeans turned their attention to the New World to obtain medical botanical lore, with scientists and laymen alike collecting information. Indeed, one of Columbus's stated purposes for his voyage to find a new route to the Indies was to obtain "spiceries of medicine ... which would be of great value in Spain" (Morison 1963:76). He had read the *Travels of Marco Polo*[12] and was going to look for the plants he had read about. In his journal he wrote, "What causes me the greatest grief in the world, when I see a thousand sorts of trees that each have their own kind of fruit ... and a thousand sort of plants, the same with flowers; and of the whole lot I only recognized this aloes, much of which I have also ordered brought aboard to bring to your Highness" (Morison 1963:79). Columbus was in fact no botanist, for he did not see even the plant he claimed to know from Marco Polo, *Aloe*.[13] Instead, he was looking at *Agave*.

For Columbus's second voyage in 1493, he had Diego Alvares Chanca, doctor for the daughter of Ferdinand and Isabella, as his official physician. Chanca reported native healing and natural history in a letter that stands as the first ethnography of the New World. Trained in the sciences of the time, he tried to identify what he saw with familiar European plants, but he mistook plants on Hispaniola for those he hoped to find, such as cinnamon and ginger (Tió 1966; Ybarra 1894, 1906). He also confused *Aloe* with *Agave*, of which he said "though not of the same kind as the one we are acquainted with in Spain" it was "a species of aloes that we doctors use" (Ybarra 1906:1016).[14]

Enter Cortés

Twenty-five years after the first voyage of Columbus, the conquistadores came to Mexico, and confusion of *Agave* and *Aloe* continued. For example, Hernán Cortés's soldiers were reported to have drank "milk of the aloe" (see MacLeish 1963:331), but because that is a drastic purge, they are more likely to have drank *pulque* or perhaps *mescal*, two alcoholic drinks made from *Agave*. Nevertheless, Cortés continued the medical ethnographical tradition of Columbus and Chanca. Writing to the Holy Roman Emperor Charles V (1519–56), Charles I of Spain, Cortés described Tenochtitlan (Mexico City) in his second letter: "There is a street set apart for the sale of herbs, where can be found every sort of root and medical herb which grows in the country. There are houses like apothecary shops where prepared medicines are sold, as well as liquids, ointments & plasters" (MacNutt 1908:257).

Cortés, eager to develop the conquered land, had made an urgent request for bringing plants from Spain: "I again pray your majesty to order a provision from the *Casa de contración* at Seville, so that no ship be allowed to sail without bringing a certain number of plants which would favor the population and prosperity of the country" (MacNutt 1908:218). Numerous plants with medicinal properties were exported from Spain, with the result that introduced plants now total 65 percent of the most frequently used herbs reported of one study in Mexico (Lozoya, Velázquez, and Flores 1988).

Cortés respected Aztec medicine and doctors: "You need not send doctors," he is said to have written to his emperor, Charles V. "The doctors here are more skilled."[15] Aztec physicians indeed could suture with hair and treat wounds with herbs unknown to European doctors. Spain was also losing medical expertise, for another event had occurred the day before Columbus set sail: Jews were expelled from Spain. Until then, medicine for monarchs, nobility, and even townspeople was almost exclusively in the hands of Jews (Granjel 1965). Moreover, the same year the Moors, who had kept alive ancient sources of medical knowledge, had been reconquered, and they took their practice underground. Some Jews and Moors had converted to Christianity, but the suppression of those whose conversion the Inquisition did not believe diminished this source of knowledge.[16] Later, some of these *conversos* went to Mexico.

The emperor Montezuma II had an experimental herbal garden at Huastepec, where medicinal herbs were brought from throughout the Aztec empire, acclimatized, and grown for use. Nothing like it existed in Europe. It is said that all that was asked of the person to whom a medicinal plant was given was that he report on the results of its use. However, other information about the herbal garden conflicts with that reputed generosity. The historian Juan de Torquemada, who was a member of the faculty of the Mexican College of Santa Cruz, wrote, "Montezuma kept a garden of medicinal herbs and . . . the court physicians experimented with them and attended the nobility. But the common people came rarely to these doctors for medical aid, not only because a fee was charged for their services, but also because the *medicinal value of herbs was common knowledge* and they could concoct remedies from their own gardens" (Emmart 1940:48; italics in original).

An herbal garden may still be seen in Cuernavaca, on the grounds where both Cortés and Mexico's ill-fated Emperor Maximilian (1864–67) had their palaces. It is used today for teaching folk healers.[17] There were also great herb gardens near Tenochtitlan and throughout the Aztec empire (Nuttall 1923). The Old World learned of Aztec medicinal plants largely from these gardens, for any medical documents that might have existed from before the Conquest were destroyed through the terrible zeal of the missionaries and conquistadores. Thus we are limited in our understanding of Aztec medicines to the information collected by European natural historians and observers such as Bernardino de Sahagún, Francisco Hernández, and others. These works were published more than a century after Cortés, which attests to the power of oral tradition.

Early Descriptions of Aztec Medicine
The first work on the natural history of the New World was published between 1535 and 1557. Serving as official chronicler of overseas Spain, Gonzalo Fernán-

dez de Oviedo y Valdés had met and interviewed Columbus and his crew in 1493. Between 1513 and 1547 he traveled in the Caribbean and Central America. In *Historia general y natural de las Indias . . .* (Oviedo y Valdés 1851–1855), he wrote about medicinal trees and plants and their properties; herbs and seeds introduced from Spain as well as those which were like those of Spain but were indigenous.

Charles I of Spain, who became the Holy Roman Emperor Charles V and was committed to a universal religious empire, had sponsored the College of Santa Cruz in Tlatelolco. It opened in 1536. There, to aid in the religious conversion of the Indians, the Franciscan friar Sahagún collected information about native beliefs, including the theory of disease and which plants cured, amassing information for a twelve-volume compendium, *Historia general de las cosas de Nueva España*, "General History of the Things of New Spain" (Sahagún 1982). He is cited today as the first researcher to use ethnographic techniques in his interviews of elderly indigenous doctors, collecting the data from their point of view in the native language, Nahuatl.

Sahagún offers us the best available window into the Aztec culture and is the source for modern scholars such as Alfredo López Austin (1970, 1980) and Bernard Ortiz de Montellano (1976, 1990). However, the compilation of the medical books inevitably reflected sixteenth-century European knowledge, distorting Aztec cosmology and disease theory (Ortiz de Montellano 1990). Moreover, only the rational, material aspects of Aztec medicine were reported (Aguirre Beltran 1947)—thus the emphasis on plants. Supernatural aspects of medicine, such as diseases caused by angry gods, were omitted because they were seen to be the "work of the devil."[18] At the same school, an herbal was written by an Aztec doctor, Martín de la Cruz, and translated into Latin by a colleague there, Juan Badiano (Gates 1939); written to impress the Europeans with its sophistication, it too neglected the Aztec theory of illness and cure (Ortiz de Montellano 1990:20).

In 1571 Philip II, the successor to Charles I in Spain, sent Hernández to study the natural history of the New World. Hernández, born into a family of *conversos*, was a scientist trained in medicine at Alcalá de Henares. Hernández was called the "Pliny of the New World" because he wrote, as did Pliny, a natural history. He used the models of his own translations of Pliny from Latin into Spanish (Hernández 1960, 1966, 1976) in his thorough descriptions of more than three thousand plants (Hernández 1959, vols. 2 and 3). Hernández's study occupied his team until 1576. Because of a series of misfortunes, Hernández's work was not published for almost a century after his death in 1587. However, he had left three or four copies of his work in Mexico, one of which was translated and edited by Father Francisco Ximénez ([1615] 1888).[19] Ximénez published for "people who live where there are neither doctors nor pharmacies," recommending healing with indigenous plants. Others who published self-help types of books (and who, according to Ximénez, plagiarized Hernández's work)[20] were Spanish doctors who came to Mexico, such as Gregorio López ([1580–89] 1672) and Juan Barrios (1607), and Mexican-born Augustín Farfán ([1592] 1944). In 1698, Father Augustín de Vetancurt wrote of the same medicinal plants in *Teatro mexicano* (Comas 1968), citing Hernández, Farfán, López, Barrios, and Ximénez.[21]

Hernández, like Sahagún, had received from his classical education a theory of medicine that interfered with his interpretation of the information that he was given in Mexico. For example, in describing *yolochichilpatli (Talauma mexicana)* he wrote, "[It] is hot and dry and resinous in nature, nevertheless the indians say that it cures erisipelas if applied as a plaster, which I cannot see how it can happen unless by dissolving the humor or by some viscous or moistening property" (Hernández 1959, 3:4). He could not understand the Aztec use because it contradicted Galen, the supreme authority on disease in Europe at that time as well as for the previous fourteen centuries. According to Galen's theory, diseases were caused by imbalanced humors and should be treated with plants having the opposite humoral properties. Erysipelas is a skin disease that in European humoral medicinal theory was thought to be caused by internal heat. Hernández (1959, 3:102) was also puzzled by the Aztec claims for *pitahaya* (*Stenocereus*, organ pipe cactus): "They say that when taken it cures swelling of the belly and hydropsia, which I do not know how it can effect that if not because some special property." *Pitahaya*, he believed, was 'cold' in the second degree and 'wet,' ineffective for a condition caused by a pathogenic 'cold' and 'wet' humor.

Although the publication of Hernández's mammoth study, which he wrote in Nahuatl and then translated into Spanish, was long delayed, some of his information reached the Old World through a physician who was likely his schoolmate at Alcalá, Nicholas Monardes. Monardes studied the medical plants he received in Seville (he never set foot outside Spain) and then wrote in 1574 the first edition of *La historia medicinal de las cosas que se troen de nuestras Indias Occidentales,* which appeared as *Joyful Newes Out of the Newe Found World* in England (1596). The book told pridefully of Columbus's accomplishments, which resulted in the riches of "many Trees, Plantes, Herbes, Rootes, Joices, Gummes, Fruites, Licours, Stones that are of great medicinall vertues" (Monardes 1596: fol. 1). His work was quickly translated into Latin, Italian, Flemish, French, and English (Rohde 1971:121), and the American herbs that he described were included in the European herbals.

Charles de L'Écluse (Clusius), who probably translated Monardes' work into Latin (Sanecki 1992), also had contact with the explorer Drake: "In the yeare (saith Clusius) 1581, the generous Knight Sir Francis Drake gave me at London certain roots . . . which in the Autumne before (having finished his voyage, wherein passing the Straights of Magellan, he had encompassed the World) he had brought with him . . . I have given them the title Drakena radix" (Gerard

[1633] 1975:1621).[22] Clusius also traveled extensively in Spain, collecting medicinal plants and their names in Spanish, seemingly the first to do so.

John Gerard (1545–1612) was a barber-surgeon, superintendent of royal gardens, advisor to Shakespeare, and herbalist for Elizabeth I and James I (Sanecki 1992). Gerard's *Herball* of 1597 was among the most famous English herbals; for example, it accompanied the Pilgrims to New England.[23] It was a tremendous compendium of some 2,050 plants, gathered from Theophrastus, Pliny, Galen, Avicenna, and all the other ancient writers combined with the sixteenth-century authors, with 1,800 illustrations that had appeared in other herbals. Each of his genera constituted a chapter giving the name of the genus; descriptions of the different species in the genus; names of places where they grew; the time of planting and maturation; names given in Greek and Latin as well as in German, Italian, Spanish, low Dutch, French, and English (and their derivations); the temperature or nature, that is, humoral properties; virtues; conditions helped by the plant; methods of administration; and finally the danger, that is, toxicity.

A revision of Gerard's work by Thomas Johnson was published in 1633 (Gerard [1633] 1975). This added another 800 plants and 700 illustrations, including more exotics found by Clusius. This work summarizes sixteenth-century knowledge augmented by discoveries from the New World; significantly, however, almost no plants of the American and Mexican West are included.

By the seventeenth century, the scientific study of medical botany in the New World had shifted to a Jesuit network of scholars such as Joseph de Acosta ([1590] 1962) and Bernabe Cobo ([1653] 1890). Other Jesuits including Georg Joseph Kamel, first in New Spain and then in the Philippines ([1688] Gicklhorn 1973:58), and Engelbert Kaempfer ([1727] 1906) in Japan expanded knowledge of medical botany by their annual reports of what transpired in their missions to their superiors, as well as through correspondence with John Ray (Gicklhorn 1973:60) and Hans Sloane (1707), both secretaries of the British Royal Society. Sloane, in describing each plant, quotes Hernández, Ximénez, and Oviedo, for he was physician in the West Indies for more than a decade.[24] It would seem that the work of the missionaries and secular scholars thus entered the corpus of medical botany.

Northwestern New Spain

In 1572 the Jesuits were allotted northwestern New Spain. For the next two centuries, these missionaries included discussions of medicinal plants in their reports. They, like missionaries even today, believed that the bodies of their spiritual charges should be aided along with their souls. The missionaries were not educated specifically in medicine, but they were trained to learn the languages of the Indians to facilitate their charges' conversion to Christianity, and in this way they studied how the Indians in their provinces healed with plants.

One was Juan de Esteyneffer, who hispanicized his name from Johann Stein-höfer because it was not then politically correct to be German. Educated as a pharmacist, he had not taken final training and vows as a priest but helped in the missionary effort as a brother co-adjutor. Sent to Mexico to nurse the old or ailing missionaries, Esteyneffer assumed the duty of helping his fellow missionaries learn curing. Because the mission field had grown more extensively than he could cover, he wrote what was considered to be a handbook of practical medicine, the *Florilegio medicinal de todas las enfermedades* ("Compendium of Medicines for All Illnesses") "for the missionaries of the Company of Jesus who work in the vineyards of the lord without recourse to doctor or pharmacy" ([1719] 1978:145). In his *Florilegio* (divided into three books), the plant, animal, and mineral remedies of western European medicine were presented together with some indigenous materials. The first described medical conditions; the second, surgical and orthopedic; the third, a "catalogue" of medicines, with recipes and directions for preparing them. This 522-page book, printed in Mexico, Madrid, and Amsterdam, was first published in 1712, then in 1719, 1735, 1755, 1887, and 1978. Most of his recommended medicines came from the classical European tradition, but he did include thirty-five plants indigenous to Mexico. The *Florilegio* is still used by healers in Mexico.

Other Jesuits who followed Esteyneffer in northwestern New Spain also noted local herbs. Juan Nentuig ([1764] 1977) reports, "Divine providence has enriched this province, destitute of doctors, surgeons or apothecaries, with such excellent medicinal products in herbs, shrubs, roots, gums, fruits, minerals and animals that you will never find such a collection in any botanical garden in Europe."[25] After listing the plants already known to medicine that were found in Sonora, Nentuig then went on to describe thirty-three new plants.

After the Jesuits were expelled by papal decree in 1767, Sonora passed to secular control. Natural histories, however, continued. Three Spanish officers—Joaquin de Amarillas (1783), Blas Antonio Pablos (1784), and Joseph Tamayo (1784)—described the plant life, including medicinal plants, of their region in their annual reports. Martín de Sessé and José Mariano Moziño (1887), commissioned in 1785 by Spain to expand knowledge of the natural history of New Spain (Engstrand 1981), went beyond the land surveyed by Hernández, to northwestern New Spain. José Longinos Martínez in his California expedition traveled and collected plants up the Pacific coast from San Blas to San Francisco (Simpson 1961; Engstrand 1981).

These scientists classified their collections by the new Linnaean *Systema naturae* and noted materials reported by the earlier missionary and secular naturalists. The plants were then entered in the official pharmacopeias of each nation. And so continued a process of medical acculturation wherein the Old World and New World exchanged knowledge of each other's healing materials. Whereas food plants of the New World enriched the diets of others for all time,

FIGURE 3 *Centaurea melitensis.*
a. Flower head. b, c. Achenes.

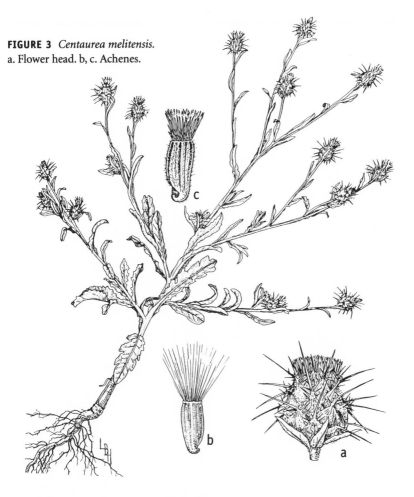

few medicinal plants from the New World have continued to be used in Europe.[26] It seems likely, by contrast, that in the New World higher status was accorded to Old World medicinal plants because they were recommended by powerful European priests. These plants, however, would have been hard for poor Indians to obtain.

Ethnic Groups of the Region and Their Medical Botanists

The area that I have labeled the American and Mexican West begins just north of the Gila River in southern Arizona, while the southern limit is immediately below the Rio Sinaloa (following Hinton 1983:315). The eastern boundary is roughly the Rio Grande, the western is the Pacific Ocean. The boundaries are determined by the colonial history of the area and also by the availability of twentieth-century ethnobotanical data: for these reasons I include the Mexican states of Baja California and exclude most of the U.S. state of California. I

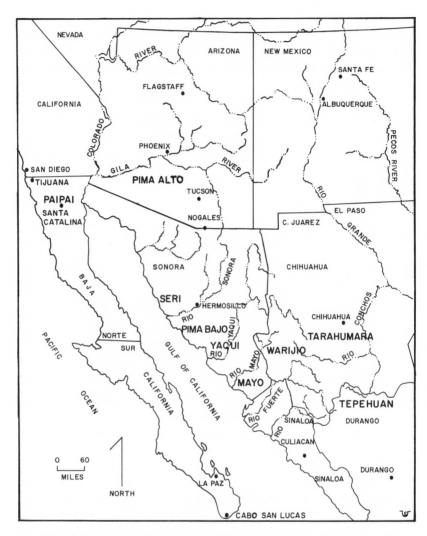

NEVADA

ARIZONA

NEW MEXICO

SANTA FE

FLAGSTAFF

CALIFORNIA

ALBUQUERQUE

RIVER

COLORADO

PECOS RIVER

PHOENIX

SAN DIEGO

GILA

RIVER

TIJUANA

PIMA ALTO

TUCSON

PAIPAI

SANTA
CATALINA

NOGALES

EL PASO

C. JUAREZ

RIO GRANDE

SONORA

CHIHUAHUA

BAJA

SERI

HERMOSILLO

RIO

CHIHUAHUA

CONCHOS

PACIFIC

NORTE

PIMA BAJO

SUR

GULF

RIO SONORA

RIO YAQUI

YAQUI

WARIJIO

TARAHUMARA

RIO

OF

CALIFORNIA

RIO MAYO

MAYO

OCEAN

CALIFORNIA

RIO FUERTE

TEPEHUAN

SINALOA

DURANGO

RIO SINALOA

CULIACAN

0   60

MILES

DURANGO

NORTH

SINALOA

DURANGO

LA PAZ

CABO SAN LUCAS

Locations of culture groups in the American and Mexican West (adapted from Spicer
and Thompson 1972:33 by Sharon Urban).

discuss the Indian and later mestizo peoples living in this area but rarely
include the Navajo and Apache, who entered Arizona and New Mexico only
shortly before Columbus, because these peoples employ many plants that are
used by Athabascans much farther north (Hrdlička 1908). Pueblo peoples are
also touched only briefly because their pharmacopoeias overlap with those of
peoples living in similar environments more directly within the study area.

   In the American and Mexican West, native peoples' first contact with the
medicine of Spain was probably in 1535, when Álvar Núñez Cabeza de Vaca

traveled through the region. He was valued as a healer, with methods that used the laying-on of hands rather than the administering of medicines. It has been suggested that the Cáhita (Yaqui and Mayo language group) word *yorim,* from the verb *yore* meaning "to heal," was applied to all Spaniards following the *entrada* of Cabeza de Vaca (Alegre 1956, 1:494).

The Spaniards conquered territories in central Mexico within a few years but had difficulty expanding to the north, where even the Aztecs had not ventured with notable success. The peoples of northern New Spain were thought to be related to the Aztec culture (Hinton 1983:315–16), including Aztec medicine, but at the time of the Spanish Conquest lived in a manner that was comparatively simple. For example, Hernández (1959, 2:290) described "the *chichimecas,* fierce, barbarous and unconquered who live not far to the north and roam the mountains and fields covering only parts of their body with animal skins," subsisting mainly on hunting. Spanish military forces, missionaries, and people looking for profit and adventure slowly pushed the ancient frontier northward (Léon-Portilla 1972:78). And as they moved, they carried their ideas of healing with them. Ultimately they achieved something of a medical conquest of the northern territories (Kay 1987). Through the centuries, the conquest of the Indians continued in cycles of domination—Spanish, Mexican, and American (Spicer 1962)—each affecting healing practices.

Of the numerous groups of native peoples recognized by the Spaniards, most in this region who have survived today are speakers of Uto-Aztecan languages. One such linguistic subgroup is the Piman, also called Tepiman, comprising the Pima Alto or Upper Pima, the Pima Bajo or Lower Pima, the Northern Tepehuan, and the Southern Tepehuan. A second is the Taracáhitan, comprising the Cáhitan Yaqui and Mayo, the Tarahumara and Warijio, and previously, the Ópatan (Miller 1983). Peoples living in the same region but not linguistically related to the Uto-Aztecans are the Seri and the Paipai. The healing plants of all these people will be discussed, for they have in common geography as well as exposure to the same missionaries for two centuries (from 1587 to 1767), and then the same governments.[27]

Most of the peoples who were placed in the Jesuit missions originally lived along permanent streams from the Gila River to the Rio Sinaloa, where availability of water and alluvial soils made agriculture possible. This concentration along the rivers facilitated the conquest and missionization of the people, resulting in an "overwhelming hispanicization that has significantly remade the cultural complexions and even influenced the physical types of these peoples" (Hinton 1983:318). The nonriverine peoples, by contrast (such as the Seri and the Tohono O'odham), were somewhat shielded from this influence, as were those who lived in the mountains and barrancas (the Tarahumara, Warijio, and Tepehuan). Similar lifeways developed among the Yaqui, Mayo, Pima Bajo, and Ópata.

The Ópata lived in central and eastern Sonora and were established in towns by the Jesuits between 1628 and 1650. They readily converted to Christianity and to Spanish patterns, ultimately disappearing as a culture by what has been called "voluntary amalgamation" (Hrdlička 1904). The medical ethnobotany of eighteenth-century Sonora, emphasizing the Ópata and including the Piman Eudeve and Pima Bajo, was described in detail in the memoirs of the Jesuits who served in their missions, including Esteyneffer ([1719] 1978), Nentuig ([1764] 1977), Ignatz Pfefferkorn ([1794–95] 1983), and Fillippe Segesser (Treutlein 1945; Segesser n.d.). Pfefferkorn ([1757] 1989:279) said that in treating illnesses among his Indians he used the work of Steinhöfer (Esteyneffer), which he claimed was "practicable in that it prescribed throughout only household remedies or well-known herbs."[28]

After the Jesuits were expelled from the New World, responsibility passed at first to the Franciscans, including Father Angel Nuñez (1777). His report to his provincial superior described plants in the upper Ópata area at Baserac. Authority then passed to secular officers of Spain. Stationed at Rio Chico was Pablos (1784), who described vegetation including medicinal plants of the land in his jurisdiction, as did Tamayo (1784) from the Valle de Tacupeta. Ópata people were also included in early twentieth-century medical observations of the Indians of the southwestern United States and northwestern Mexico (Hrdlička 1908).

North of the Ópata are the Pima Alto, who call themselves Ó'odham, "we the people." Those living near the Gila River are usually referred to as the Pima while those farther south, formerly called the Papago, are officially known today as the Tohono O'odham. Their lifestyles have derived from their environment, for nearly the whole of the Pimería Alta lies within portions of the Sonoran Desert, mainly dry and forbidding land (Fontana 1983). The Pima are believed to be connected to the ancient Hohokam culture that once spread throughout the same area; some archaeologists theorize that the canal-building Hohokam might have originally come up from Mexico.[29] Pima medicinal plant use consists almost entirely of plants indigenous to this area (Curtin 1949); plant use by the Tohono O'odham is somewhat different, reflecting the varied landscape of the desert (Hrdlička 1908; Nabhan 1985).

The Pima Bajo people, known in colonial times as Névome (when they included Jova and Eudeve), were served by Nentuig and Segesser, who recorded their medical ethnobotany. After the Jesuit expulsion, Tamayo (1784) reported from the Pima Bajo settlement of Arivechi: his *relación* listed plants and their uses. Modern scholars have also studied the medicinal plants of the Pima Bajo, including the Mountain Pima (Pennington 1973, 1980; Laferrière, Weber, and Kohlhepp 1991).

Well to the southeast of the Pima Bajo, the Tepehuan live in the present-day state of Chihuahua, in the southern canyon or rolling upland territory. Jesuit missions were established at Nabogame and Baborigame shortly after 1700, introducing many European plants and animals. In their traditional medicine the Tepehuan used plants of at least fifty-six families (Pennington 1969).

The Tarahumara are neighbors to the north of the Tepehuan. They live in some of the most rugged terrain in North America, in the uplands and canyon country of southwestern Chihuahua, where they retreated from Spanish forces and moved into territory once dominated by other tribes. From the first contact in 1610, Jesuits have been the most lasting influence on their lifestyle (Pennington 1983a). The Tarahumara grow the usual corn, beans, and squash for a marginal subsistence that is gravely threatened whenever there is drought. In contrast to the Ópata, the Tarahumara have remained culturally distinct despite the efforts of various groups, from the Mexican government to Catholic and Protestant missions. Medicinal plant use by the Tarahumara has been described by many ethnographers,[30] and the Tarahumara are famed as collectors of medicinal plants that are sold in markets throughout Mexico.

The Mayo lived in the southernmost part of the region in two different zones, the thorn forest desert areas and the Fuerte River valleys, but have since been pushed out of the fertile valleys by Mexican farmers. In the seventeenth century the Mayo welcomed Christianization (Pérez de Ribas [1645] 1944), although their acceptance was not so complete as the missionary may have wished; like their neighbors the Yaqui, the Mayo include many native elements in their religious practices. For example, dancers in the Easter ceremonies of both the Yaqui and the Mayo often are people who have made a vow *(manda)* to participate after they have been cured of a serious illness.[31]

The Yaqui have been called one of the "enduring peoples" of the world (Spicer 1980). After successfully resisting the conquistador Nuño Beltrán de Guzmán in 1533, they had little contact with European culture until Jesuit missionaries in the seventeenth century induced them to leave their scattered *rancherias* and settle in eight villages in the Rio Yaqui valley. Their relations with Mexicans were always rebellious, culminating in their deportation to Oaxaca at the end of the nineteenth century by the orders of Mexico's dictator Porfirio Díaz. At different times groups of Yaqui migrated to Arizona, seeking freedom. More remain in Sonora, living in the southern part as well as in and around Ciudad Obregón. There are three settlements of Yaquis in Tucson, as well as one in Marana north of Tucson, another in Guadalupe near Phoenix, and another in Scottsdale. Participation in Yaqui Easter ceremonies is an important way in which thanks are given to God for cure from an illness. In other ways, their healing practices resemble those of the Mexicans. In 1783, a secular deputy, Joaquín de Amarillas, reported on the ethnobotany of the Yaqui. Despite the difficulty of studying the far-flung Yaqui, more recent researchers have described their medical practices and included the names of the plants that they used.[32]

The Warijio (also Guaríjio, Warihio, and Varihio), unlike the other groups in the American and Mexican West, have had little attention from ethnographers since Andres Pérez de Ribas's (1944) report of 1645 on the activities of Jesuit missionaries. After the Warijio were "rediscovered" in 1930, the botanist Howard Scott Gentry (1942, 1962) believed that their culture had derived from their barranca habitat and, as hunters and gatherers, their need for short local migrations. At the time that this botanical study was conducted, the Warijio lived isolated from much of Mexican culture in the valleys and barrancas of the upper Rio Mayo; but as Mexicans moved into the area, the Warijio adapted the Mexican lifestyle, adding agriculture to their gathering by growing maize, squash, sugarcane, amaranth, chile, onions, tobacco, and cotton, as well as salvia. Like their neighbors the Tarahumara, they speak a Cáhitan language of the Uto-Aztecan family, and in other ways the cultures are also similar.

The states of Baja California Norte and Baja California Sur have histories of medicinal plant use that differ from each other. Father Eusebio Francisco Kino, a mapmaker and the most famous of the Jesuit missionaries to northwestern New Spain, proved the peninsular nature of Baja California. With the introduction of the Jesuits to the area came more physical descriptions. Father Adamo Gilg (1692; see Schuler 1973:498–505), Juan Maria Salvatierra (Clavijero [1786] 1937), and the familiar Juan de Esteyneffer (Dunne 1940) were among Kino's companions who wrote about Baja California. Father Miguel del Barco ([1767] 1973) wrote a natural history of the area, describing medicinal uses of some plants. Father Francisco Xavier Clavijero, who served in Baja California for thirty years, also wrote an extensive history of Baja California ([1786] 1937), including discussions of native and introduced medicinal plants, after he was sent to Italy when all Jesuits were deported from Mexico. He also wrote *History of Mexico* (1787), which included a discussion of medicine.[33] Recently, researchers have interviewed herbalists in Baja California Norte and elderly informants at the southernmost part of Baja California Sur regarding medicinal plants and collections in the Municipio de los Cabos and part of the Municipio de la Paz.[34]

Unlike most other native groups reported herein, who are Uto-Aztecans, the Paipai are Yuman-speaking. They live in the northernmost part of Baja California Norte in an upland desert area characterized by creosote bush, juniper, manzanita, prickly pear, cholla, and agave. Their first significant European contacts were with Dominican missionaries who established a mission at Santa Catarina in 1798. In 1840 the Paipai burned the mission and drove out the missionaries, and since that time they have never been under the directive control of either European missionaries or Mexican civil administrators. Thus they are a population whose use of medicinal plants is especially interesting. One study (Owen 1963) shows that more than half of the medicinal plants used by the inhabitants of Santa Catarina were of European-Mexican origin: of sixty-seven plant species, seven were introduced to Baja California after European contact, four were recently borrowed from Mexico, fourteen have similar usages in other Latin American areas, and twelve lack Paipai names (1963:342), which suggests they also were borrowed.

The Seri, also non-Uto-Aztecans and speaking a Hokan language related to Yuman, live across the Sea of Cortez in an extremely arid environment along the Sonoran coast of the Gulf of California and Isla Tiburón. A hunting and gathering people, they had little friendly contact with the Spanish and were always an enclave of resistance to Spanish encroachment. They remained isolated from European influence as well as later Mexican culture, with their culture persisting vigorously into the middle of the twentieth century. They collect medicinal plants from the desert and have less Mexican influence on their herb choices. Seri ethnobotany has been comprehensively described (Felger and Moser 1985), as has their obstetrical ethnobotany (Moser 1982).

The conquest of Mexico began a new people—mestizos—at first with Spanish fathers and indigenous mothers. The conquistadores were the first Spaniards to explore the northern lands, which have since become the American and Mexican West. A report of the inspection and inventory of supplies held by don Juan de Oñate for the conquest of New Mexico in 1596 included a detailed list of medicines. Of these medicines, only *zarzaparilla* (*Smilax,* sarsaparilla) and *estafiate* (*Artemisia,* mugwort) were indigenous to Mexico, and even these genera were represented in the European pharmacopoeia.[35] As industries such as mining for silver and cattle ranching developed, mestizos began to fill the northwestern territories. They sent home for familiar remedies to Mexico, Spain, and elsewhere. And although plant medicines of the New World were sent to Europe, apparently none came from what is now northwestern Mexico.

Mexico achieved independence from Spain in 1821. Next came war with the United States, with the eventual loss of half of Mexico's territory, including the present states of Texas, New Mexico, Arizona, California—the American West. The Treaty of Guadalupe Hidalgo (1848) and the Gadsden Purchase (1853) that followed the Mexican War established the present national boundaries, transferring many Mexicans to the United States and dividing the O'odham people on both sides of the border. Yet this political division of this land is brief, relative to the history of the land.

The People and Their Medicinal Plants Today

The plants described by the various ethnographers of Indian groups cited above are all employed today as medicines by traditional Mexicans, those living in Mexico as well as those in the United States, who are called Mexican Americans. Healing with plants varies by regions and by particular groups. For example, regional differences in plant use are visible among *norteños,* Baja California

Mexicans, Sonorans, and Chihuahuans, who also exhibit generational and economic differences. Similarly, Mexican Americans in New Mexico, Arizona, and Colorado also vary in the extent to which they use herbs, related to such characteristics as generation removed from Mexico, age, ideology, and economics. Some Mexican Americans cross the border regularly to purchase herbal remedies, while others can obtain them locally. Every city with a Mexican American population has a *yerberia* somewhere, and near the border there are herb stores or stalls in markets, even in smaller settlements. In Tucson there are several *botanicas* or *yerberias:* one with a large inventory of traditional Mexican remedies is owned and operated by a Cuban man who obtains his stock from various wholesale houses in Los Angeles, San Diego, Colorado, and also Nogales, Sonora. The same wholesale houses provide medicinal plants to supermarkets: one inventory sheet listed 221 items.[36]

Traditional Mexicans, who may live on either side of the border, may also grow medicinal plants in their kitchen gardens and collect others in the chaparral, fields, ditch banks, and mountains. New fashions in medicines bring certain plants to favor. For example, some medicinal plants are touted for *gordura* (obesity), diabetes, and elevated blood pressure, conditions that popular culture now emphasizes. I have found that everyone knows a certain neighbor who grows medicinal plants in his or her garden—but usually before this knowledge is shared, *confianza* (trust) must first be established. It took months after I first became acquainted with one such neighbor to learn that marijuana is an important ingredient in her liniment for *artritis* (arthritis).

Education, income, and the availability of traditional healers and biomedical facilities all relate to the persistence of traditions of healing with herbs, but not necessarily in expected ways. In Mexico, data from a study conducted by the Mexican Social Security Institute and Community Cooperation Program (Lozoya, Velázquez, and Flores 1988) show that the Mexican West differs from the rest of Mexico in many social aspects that would affect healing. For example, there are differences in the ratio of traditional healers (who use herbs) to biomedical practitioners (who do not use herbs) and in the percentage of persons more than fifteen years of age who are nonliterate. Availability of different types of healers would be expected to affect use, as would accessibility. Literacy would also be reflected in the amount of income available for health care and in a tendency to depreciate traditional healing. However, the actual data run contrary to the expectations of common wisdom.

For example, the state of Chiapas, in the south of Mexico, has a 2:1 ratio of traditional healers to biomedical care and an illiteracy rate of 28.90; by contrast, Baja California has a 6:1 ratio of traditional healers to biomedical care and an illiteracy rate of 6.10, while the national averages are 4:1 and 16.45 (Lozoya, Velázquez, and Flores 1988:55). Despite the much lower percentage of nonliterate persons in the states of the Mexican West, the much higher ratio of traditional healers to physicians may explain the persistence of herbal treatment. Further, these traditional healers are comparatively young (Lozoya 1990). Xavier Lozoya (1990:77) has hypothesized that traditional medicine has been revitalized in a renewed role in states such as Baja California Norte and Sonora because large-scale population movements have occurred without a corresponding expansion of biomedical services.

The Lozoya study also surveyed the plants used for medicine in all the states of Mexico. After correction of various spellings about 1,950 species were reported in the survey: 35 percent of these were indigenous, 65 percent introduced, mainly from the Old World. Of the 140 species commonly used throughout Mexico, the most frequently appearing herbs were mint, rue, chamomile, corn silk, basil, artemisia, chenopodium, catnip, and aloe. Only three of these nine are indigenous to Mexico. For Baja California Norte, Sonora, and Chihuahua, the most common were mint, manzanilla, artemisia, corn silk, creosote bush, horsetail, eucalyptus, damiana, aloe, and basil, only half of which are indigenous.

Both the ecology and the history of the American and Mexican West have largely determined which plants have been used for healing. Isolated Indian groups once used more of the plants that grew in the immediate environment, some indigenous and others learned in trading with other peoples. Today the majority of people living in Mexico are mestizos whose medicinal plant use comes from the pre-Conquest and colonial periods; yet the American and Mexican West has become increasingly multi-ethnic, populated with Mexicans, Spaniards, English, Anglo Americans, Irish, African Americans, Chinese, Japanese, Thai, and others, all bringing their traditional uses of plants as medicines. All these peoples have accepted new remedies when the old could not be obtained or when they learned of native plant medicines. This is one of the many universal themes in medicinal plant use.

2 ▪ Plants, Their Names,

and Their Actions

The history of use of specific plants is often reflected in their names. Both common and scientific names may suggest a plant's appearance, or common names in the various languages may indicate use, including healing. What is more, a given common name is commonly applied to more than one species in a genus of plants and may even be applied to a plant in a different genus, while a single species may have multiple common names even within the same language.[1] Bearing these problems of naming in mind, I have documented one hundred genera of the plants most commonly used as medicines (see part 2) by the peoples of the southwestern United States and northwestern Mexico, plants that produce phytochemicals capable of preventing or curing an illness. The traditional plant pharmacopoeias evolved to their present forms in each society through a unique combination of ecological, historical, and cultural factors.

The American and Mexican West

Much of the American and Mexican West is characterized as desert. However, within the area there is diversity: biotic communities of forest and woodlands, scrublands, grasslands, desertlands, and wetlands (Brown 1982). Each region favors a certain type of vegetation (Rzedowski and Equihua 1987) where specific medicinal plants can be found.

Yet although peoples such as the Tepehuan and the Tarahumara live in rather similar environments (in uplands and in canyons), few plants are used as medicines by both groups—these are in almost all instances the most widely distributed medicinal plants in western and southern Chihuahua (Pennington 1969:333).[2] This observation brings up the question of what determines which plants a people employ as medicines.

Can ethnicity explain the differential selection of medicinal plants? If this is the case, then speakers of similar languages would be assumed to share culture and to use plants in the same way. The Tarahumara, Warijio, and Mayo speak related languages labeled Taracáhitan (Miller 1983), while Tepehuan belongs to a distant Uto-Aztecan language group, Tepiman. However, speakers of Taracáhitan do not seem to share use of more medicinal plants with other speakers of Taracáhitan than they do with speakers of Tepiman (see table, pages 38–41).

An alternative hypothesis, perhaps more valid, is that similarity in the biotic region can explain variation in the selection of medicinal plants. The ethnic groups in the table have been placed so that plants used by peoples of the early times (Old World and Aztec) can be seen first, then those used by speakers of Tepiman languages (Pima Alto, Pima Bajo, Tepehuan, and Mountain Pima), followed by those used by Taracáhitan speakers (Tarahumara, Warijio, Mayo, and Yaqui), Baja Californians, and the Seri. Mexican Americans are placed at the end for greatest visibility since it is the development of their plant pharmacopoeia that is a goal of this study. Mexican Americans appear often to have obtained their plant pharmacopoeia from two ancient traditions: classical European and pre-Conquest Indian (chapter 1). Within the American and Mexican West, the biotic region that favored the growth of a particular plant then explained choice in most instances.

Plant Names

Plant names have histories, but their nomenclature has not been continuous. This makes it difficult to be sure whether the label that is used today is the same that was used in the past. Plants that were introduced from Europe to the American and Mexican West carried their European names. Some of the plants had been named for certain people: Pliny (1938, 7:153) stated, "It was one of the ambitions of the past to give one's name to a plant [which] as we shall point out was done by kings," exemplified by the plant genus *Euphorbia*. Plant names such as *Achillea, Centaurea,* and *Nymphae* came from mythical figures. When Linnaeus standardized Latin binomial nomenclature in 1754 (binomials had been used to differentiate what were believed to be related species by both Pliny and Dioscorides), he commonly chose the names by which the plants had been known for the previous centuries. In the Linnaean system, the common name in Greek or Latin often became part of the official botanical binomial (for example, *Artemisia* is the genus name, coming from the Greek, while descriptors differentiating individual species include *tridentata,* "three-toothed," referring to the leaves; *ludoviciana,* to Louisiana, where the species was first described; *frigida,* to cold regions, etc).

By comparison, the Nahuatl system had built into the name aspects of the

Medicinal Plant Genera in the American and Mexican West, by Culture

Genus	OW	Azt	18c	PimA	PimB	Tepe	MtPi	Tara	Wari	Mayo	Yaqu	PaiP	BC	Seri	MexA
Acacia	+	+	+		+	+	+	+	+	+	+		+	+	+
Achillea	+	+													+
Agastache		+					+	+							+
Agave		+	+		+	+	+	+	+	+		+	+		+
Allium	+		+				+				+				+
Aloe	+	+	+			+	+	+	+	+					+
Ambrosia	+		+	+	+	+	+	+		+	+		+	+	+
Anemopsis			+	+						+		+	+	+	+
Arctostaphylos	+	+			+	+	+					+		+	
Argemone	+	+	+		+	+		+	+	+			+		+
Aristolochia	+	+	+		+				+	+	+	+	+	+	+
Artemisia	+	+	+		+	+	+				+	+			+
Arracacia		+					+						+		
Asclepias	+	+	+	+		+	+	+	+	+	+	+	+	+	+
Baccharis	+	+	+				+	+		+			+	+	+
Bocconia		+				+									
Buddleia		+			+	+	+	+	+	+			+		+
Bursera		+	+		+			+	+	+			+	+	+
Caesalpinia		+	+		+			+	+	+					
Cannabis	+		+							+		+			+
Capsicum		+	+	+	+			+		+					+
Carnegiea										+				+	
Casimiroa		+	+			+	+						+		
Cassia	+	+	+		+									+	+

Cereus	Chenopodium	Citrus	Datura	Ephedra	Equisetum	Eryngium	Euphorbia	Eysenhardtia	Gnaphalium	Guaiacum	Guazuma	Gutierrezia	Haematoxylon	Haplopappus	Heterotheca	Hintonia	Ibervillea	Jacquinia	Jatropha	Juniperus	Karwinskia	Kohleria	Krameria	Lantana	Larrea	Ligusticum	Lippia
+	+	+		+	+	+	+	+	+	+	+	+	+	+	+	+			+	+		+	+		+	+	+
+			+				+			+			+				+	+	+				+	+	+		+
+	+		+		+		+		+					+	+				+		+			+	+	+
		+	+				+	+						+							+				+	
+			+				+			+			+			+		+	+				+			
+		+	+				+	+		+	+		+			+	+	+	+			+	+	+	+		+
			+					+	+	+			+			+	+		+	+		+			+		+
+	+	+	+		+	+	+	+					+			+				+	+			+	+	+	
	+	+	+	+	+									+	+			+			+				+		
	+	+	+				+	+	+			+		+			+				+	+			+	+	
	+	+	+				+				+			+					+		+	+			+	+	
+			+	+				+							+									+	+		
+	+	+	+			+	+	+		+			+				+		+	+	+			+		+	+
	+	+	+				+	+	+	+	+			+				+	+		+		+			+	+
	+	+	+	+	+	+				+	+									+				+		+	+

Table—continued

Genus	OW	Azt	18c	PimA	PimB	Tepe	MtPi	Tara	Wari	Mayo	Yaqu	PaiP	BC	Seri	MexA
Lysiloma		+	+		+	+		+	+	+					+
Malva	+	+	+			+	+	+							+
Mammillaria		+		+	+			+						+	
Mascagnia					+			+					+	+	
Matricaria	+	+	+			+	+	+	+	+					+
Mentha	+	+	+					+	+	+			+		+
Nicotiana		+	+		+			+	+			+	+		+
Ocimum	+	+	+									+	+		+
Opuntia		+	+	+			+	+		+	+	+		+	+
Perezia		+	+			+	+	+	+	+	+				+
Persea		+	+			+		+							+
Phaseolus	+	+	+	+	+		+	+				+			+
Phoradendron	+	+	+	+		+	+	+						+	
Physalis		+	+				+	+							+
Pinus	+	+	+			+		+				+			+
Pithecellobium		+			+								+	+	
Plantago	+	+	+	+		+	+	+				+	+		+
Plumeria		+	+		+	+		+							
Populus	+	+		+		+	+			+			+		+
Porophyllum		+	+	+	+		+	+	+		+	+	+	+	+
Prosopis		+					+	+	+	+		+	+	+	+
Psacalium		+			+	+	+	+		+					+
Punica	+		+												+
Quercus	+	+	+			+									+

Abbreviations: OW = Old World; Azt = Aztec; 18c = eighteenth-century observers; PimA = Pima Alto, including Tohono O'odham; PimB = Pima Bajo, including Ópata and their descendants; Tepe = Tepehuan; MtPi = Mountain Pima; Tara = Tarahumara; Wari = Warijío; Yaqu = Yaqui; PaiP = Paipai; BC = other Baja Californians; MexA = Mexican Americans.

plant such as life form (tree, vine), habitat (riverine, montane), sensory qualities (appearance, taste, tactile feel, odor, sound), and use (edible, medicinal,
ornamental, economic),[3] but this system was overridden after the Spanish
Conquest. Francisco Hernández often adopted the plant name in Nahuatl as
the first part of a binomial in his plant discussions. For example, *cihuapatli* was
the generic name for plants used to treat women's problems (the prefix *cihua-*
refers to women, the suffix *-patli* to medicinals): *cihuapatli mayor* was the
largest species; *cihuapatli yyauhtlino* had leaves that resembled the plant
yyauhtli; and *cihuapatli anodino* (anodyne) was used for pain.

Bernardino de Sahagún and Hernández recorded Aztec plant names in Nahuatl; Hernández also recorded some Tarascan and Huastecan names. But New
World species usually lost these names later. Rather than continuing Aztec or
other indigenous names, in the Linnaean nomenclature the species were often
named for famous European botanists in latinized form (Stearn 1990). Thus
cocoxochitl[4]—the Aztec name incorporating its medicinal use (to relieve the
pain of urinary tract disorders) and its beauty (*-xochitl,* flower)—was re-named
Dahlia after Linnaeus's pupil Andreas Dahl. This was, as Jamaica Kincaid (1992:
159) sorrowfully noted, an act of conquest: "These new plants from far away,
like the people far away, had no history, no names, and so they could be given
names. . . . This naming of things is so crucial to possession" (Kincaid 1992:159).

Identification of Aztec Plants

Providing botanical names of Aztec plants is difficult. Before Linnaeus, the taxonomy of plants was not built on the morphology of the flower, as it is at present. Rather, the leaves, which we now know to be notorious for their variation,
and the roots were the criteria for classifying.

Hernández believed that he recognized some plants in Pliny's *Natural History* as the same as ones he knew in Spain. He was often accurate, as in *saúco*
(elder), assigned today to the genus *Sambucus:* "what the Mexicans here call
xúmetl is the same as *saúco* of our country" (Hernández 1959, 3:218). The
medicinal use of a *Sambucus* species was already ancient when the Egyptian
medical papyri were inscribed some three thousand years before. Sometimes
Hernández saw a likeness that is inexplicable, as when he compared *chichi-
tzompotónic* (pearly everlasting is the English common name) to *gordolobo:* "It
is an herb very like the gordolobo of our land and can be classified among
those species, with large, soft, white, hairy leaves" (Hernández 1959, 3:87). Hernández's description of the leaves fits those of *Verbascum thapsus* (mullein),
known in Europe since ancient times as *gordolobo;* however, the leaves of *Gnaphalium conoideum,* which subsequent botanists have identified as this Mexican *gordolobo,* are soft and white but not large. The flowers of *Verbascum* also
are quite different from the many-petaled flowers of *Gnaphalium.* Regardless,
in the Mexican pharmacopoeia, *Gnaphalium* was given the common name of

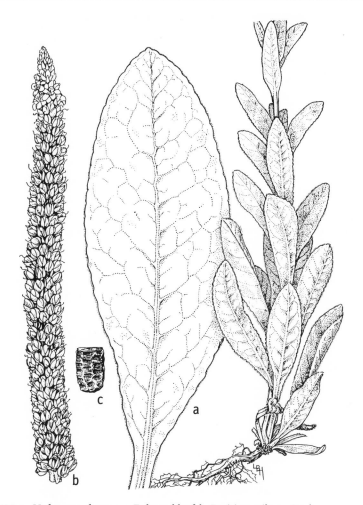

FIGURE 4 *Verbascum thapsus.* a. Enlarged leaf. b. Fruiting spike. c. Seed.

gordolobo mejicano. Although *Verbascum thapsus* is used by many North American Indians living in areas where it became endemic after its introduction from Europe, in the American and Mexican West only the Navajo, the Hopi, and Mexican Americans in Colorado are reported to employ it.

To complicate matters further, *Senecio,* called *iztacatzóyatl* by Hernández and *calancapatli* in the American and Mexican West, has slender and hairy leaves very similar to those of *Gnaphalium.* Administration of wrongly collected *Senecio* instead of *Gnaphalium* as *gordolobo* for unknown periods of time has resulted in poisoning (Huxtable 1980; Kay 1994).

Aztec names of plants, their spelling, and the botanical identifications varied among Hernández (1942; 1959, vols. 2 and 3), Francisco Ximénez ([1615]

1888), Maximino Martínez (1969), and Bernard Ortiz de Montellano (1990). The best solution was to follow the work of Javier Valdés and Hilda Flores (1984), who undertook the Herculean task of compiling botanical identifications for the seventh volume of Hernández's opus. For 1,014 of 3,076 plants, a possible identification was made. Drawings that illustrated some of Hernández's work, for example of the different species of *Opuntia* (prickly pear and cholla), also helped.[5]

Plant Nomenclature Problems and Decisions

Providing a correct botanical identification in contemporary nomenclature of the plants described in eighteenth-century books and manuscripts is problematic. Typically, the historian will consult the most authoritative dictionaries or folklore compendia that he or she can find. However, in this way errors are made and perpetuated. One needs specimens, but herbariums do not commonly contain material from the sixteenth through eighteenth centuries. An example of recent errors is the new translation of Juan Nentuig's *Rudo Ensayo*, subtitled *A Description of Sonora and Arizona in 1764* (1980), which gives scientific names to species that cannot be found in Sonora. In addition, the eighteenth-century writers had varying degrees of familiarity with the works of earlier botanists and had different degrees of linguistic skill. Thus, Fillippe Segesser in the eighteenth century wrote in German about *cohobe* and *presil;* from his descriptions of their uses it seems likely that he meant jojoba and brazilwood, respectively. Further, he confused aloes and agaves, an error continued since the time of Columbus, 250 years before: "The growth called *aloe* in Ingolstadt is called *magei* here. In several years time it produces one stalk, as I saw in the bishops' garden at Eichenstädt. Juice extracted from the maguey is refreshing to those who have inflammatory fevers" (Treutlein 1945:180).

Identifying the Sonora plants, especially those known only by their Ópata names, is particularly difficult. Plants known as "*yerba de*" are hard to fix because the same label is given to different species—indeed, to plants in different families. Those called *yerba del oso* may be one of many favorite foods of bears. The various *yerba del pasmo* are all used to treat *pasmo,* an illness no longer recognized by official medicine. A *yerba colorada* (red herb) may be so named because it has red roots, or red leaves, or produces a red brew. The same plant may be given different names for its life stage or part, such as *flor de, raiz de, semilla de, fruta de* (flower, root, seed, or fruit of).

Undeniably, the names of plants always pose a quandary (Croom 1983; Huxtable 1990). In part 2 of this book, plants are first listed by the Latin botanical name of the genus and its family. Then each species is given its botanical name, followed by common names in indigenous languages, Spanish, and English including alternative pronunciations or spellings of the same names as necessary. As you will see, the same species may be given different common names, and different species may be given the same common name.

Plants tend to have local common names, often inspired by what the plant is thought to resemble (e.g., the Mayo common name for *Rhynchosia* means bird's eye; the Tepehuan, thrush's eye; the Aztec, crab's eye) or its medicinal purpose (e.g., fever plant). These common names might be literal translations of the Spanish, which originally might have been translations from the indigenous terms. One may postulate that when the names of a species are distinctly different for each language or dialect, plant use was not learned by contact but was already known for a long time. This is exemplified by the fact that mesquite *(Prosopis)*, although named from the Nahuatl *mizquitl*, has numerous local names.

This practice, of course, is not unique: a plant may have a different common name wherever it grows. The great Mexican botanist Martínez, in his comprehensive catalogue of common and scientific names of Mexican plants (1979), listed thirteen different species of plants named *hierba del pasmo*. The Martínez text abounds with such naming: thirteen *hierba anis* (anise) and eight each of *hierba del indio* (Indian herb) and *hierba del sapo* (toad's herb).

As another example, in Mexico there are thirty common names for *Acacia farnesiana*. Plant names may be dialectical or phonological variations; *Acacia* names in Cáhitan languages are *mokowí* in Tarahumara, *coo-ca* or *kuká* in Warijio, *kukka* in Mayo, and *cuca* in Yaqui, four related languages. A further problem is orthographic. In Spanish, the common name of *Acacia farnesiana* is written *guisache, huichace, huicachin,* and *quisache,* but most commonly *huisache,* deriving from the Aztec name, *huixachi.* Many of acacia's names refer to the thorns, as in its Latin botanical name (from the Greek *akis,* a sharp point) and the *hui-* prefix in Nahuatl, which has carried over into Spanish. It is called *vinorama* or *binorama* in Baja California and Sonora. That name is a hybrid, with *bino-* from the Cáhitan *binolo* or acacia and *rama,* Spanish for branch (Sobarzo 1966:43).

Mexico's IMEPLAN-COPLAMAR study found an average of four synonyms or phonetic differences per plant (Lozoya, Velázquez, and Flores 1988:73). It must also be considered that names might not have been recorded with linguistic precision.

Moreover, there are other complications: the official Latin name may have been changed from the time that the plant was first identified. Botanists change Latin plant designations as they explore plant relationships; they rename a species, assign it to a different genus, or change the family label. For example, botanists have changed to the family name of Asteraceae for Compositae, Lamiaceae for Labiatae, Fabaceae for Leguminosae, Apiaceae for Umbelliferae, Brassicaceae for Cruciferae, and Poaceae for Gramineae (Moerman 1986:xiv).

The genus may be assigned to a different family—for example, *Simmondsia chinensis* (jojoba), formerly classified as a Buxaceae (box family), now has its own family, Simmondsiaceae—while the family Cannabinaceae has now been merged with Moraceae. The name of the genus is also subject to change. *Cacalia* is now more properly *Psacalium* (although other botanists believe the genera to be different), *Coutarea* is *Hintonia*, *Perezia* is *Acurtia*, and so forth.

What then is the best method to identify these plants? The procedure that I used for deciding identification in this study was as follows: for medicinal plants listed in this century, and if the plant usage was described by a botanist, I used his or her identification. For many of these plants, the botanists deposited a voucher specimen in a designated herbarium. But the medicinal plant that had not been identified by a qualified botanist offered a challenge. When in doubt, I compared information among field guides and other descriptive references from the sixteenth through the twentieth centuries. However, these references are not consistently useful; herbalists often give varying descriptions and names to the individual plant species they use. Sometimes dictionaries may also be helpful in providing etymological information. For this study, I consulted every useful reference of the above types for each medicinal plant.[6]

I have a collection of medicinal plants made through the years (and continually added to), either transplanted to my garden or carefully dried. Those given to me from local kitchen gardens are identified under the direction of Charles T. Mason, curator emeritus of the University of Arizona Herbarium. However, many of the herbs that I have purchased at various stores are represented only by those parts used medicinally, such as cross sections of roots or crumpled leaves that are impossible to identify. They can give little help; in fact, the labels are on occasion seriously incorrect (see entries on *Gnaphalium* and *Senecio* in part 2).

To identify plants from the ethnomedical reports of the eighteenth century, I turned to related data from contemporary ethnobotanical research:[7] if the eighteenth-century writer described the plant and gave it a name by which a medicinal fitting the description is called today, I have listed the tentative identification. The plants chosen for discussion are primarily indigenous medicinal plants, but because by the eighteenth century many species introduced from Europe had become naturalized in Sonora, a few of these plants are included. These plants are used throughout Mexico, and they are known everywhere by their Spanish common names.

Their naming may be often confusing and capricious, but the usage of plants for medicines has been continuous. Aztec medicinal plants described by Sahagún and Hernández in the sixteenth century, plants found late in the eighteenth century by Martín de Sessé and José Mariano Moziño's scientific expeditions, California plants discovered by José Longinos Martínez in his expedition up the Pacific coast—these were still to be found in the official Mexican pharmacopoeia at the beginning of this century and in the herb stores today. Thus, the plants recorded by the missionaries and the seculars earlier in the eighteenth century appear to be still known and used.

Selection of Plants

I chose plants in one hundred genera for this book. Almost all are native to the American and Mexican West; interestingly, the few exceptions are among the most commonly used, not only in this region but elsewhere in Mexico. To compare medicinal plant use for the rest of Mexico, I again referred to the comprehensive study of traditional therapy conducted by the Mexican Social Security Institute (Lozoya, Velázquez, and Flores 1988:69). The number of medicinal plants reported for all the states in Mexico totaled 5,773. The authors reduced these to 1,950 by grouping duplications, synonyms, and phonetic variants of names (on average, four different names were in use for each plant). The Mexican states of the west reported proportionately fewer plants than the southern states (Lozoya, Velázquez, and Flores 1988:72): for Baja California Norte, 47 medicinal plants are reported; for Baja California Sur, 54; for Chihuahua, 154; and for Sonora, 109. By comparison, 542 medicinal plants are reported for Oaxaca, a southern and tropical state.

Approximately 35 percent of the species used throughout Mexico are indigenous, with 65 percent introduced after European conquest. In every state the most commonly reported plants are *yerba buena* (*Mentha piperita,* peppermint and *Mentha spicata,* spearmint), *manzanilla* (*Matricaria recutita,* chamomile), *albahaca* (*Ocimum basilicum,* basil), and *poleo* (*Mentha pulegium,* pennyroyal)—all brought to Mexico with the Spanish culture, and all used for gastrointestinal disorders. The success of the introduced plants for domestic healing may have something to do with the nature of the plants: the introduced plants tend to be less toxic. However, they often need cultivation and also require more moisture than occurs naturally in arid regions.

By contrast, plants in arid regions have evolved so that they can flourish, protected against predators. Structural protection comes from thorns that can keep animals away, chemical protection from the biosynthesis of certain components that are toxic to the animal, bird, or insect that would destroy it. Indeed, "plants growing in these regions have high contents of alkaloids and tannins in comparison to their counterparts growing in non-arid conditions" (Tewari 1979:186). Indigenous plants that evolution selected to survive in the hostile environment of the desert, such as cacti, have powerful compounds stored in their roots that can prove dangerous in lay hands. Indigenous plants that are most commonly used as medicines in Mexico include *pelos de elote* (*Zea mays,* corn silk) for diuresis, *estafiate* (*Artemisia ludoviciana,* mugwort) for intestinal spasm, and *epazote* (*Chenopodium ambrosioides,* wormseed) as a vermifuge. These are usually cultivated plants.

In the northwestern Mexican states I found about 224 plants that were candidates for this study, including all those just mentioned. I organized these in a data assemblage that contains 154 genera, a few represented by a single species. I decided to choose the 100 genera used by four or more different peoples[8] and to combine species by genus. Merging species is problematic. I looked at various factors when deciding whether to report species together by genus. Often the exact species of samples that I brought for identification could not be precisely determined. For example, the many *Gnaphalium* species that can be found in the American and Mexican West closely resemble each other. Differences among *Datura* species depend on locale; they are all called *toloache* and used for the same purposes. *Achillea lanulosa* is native while *A. millefolium* was introduced from Europe; they differ only in chromosome number, and both are used as medicines. Moreover, the ancient natural historians from Pliny through Hernández and Gerard listed various names for the medicinal plants they described and combined their species by the generic categories of their time. Classifying things together is cognitively universal, albeit unacceptable to botanists.

How Plants Treat Illness

Medicinal plants are defined as those which produce one or more active constituents capable of preventing or curing an illness. In fact, these plants contain numerous chemical compounds considered to be secondary metabolites: natural chemicals that are not required for primary life support of the plant but serve other purposes. Some may be allelochemicals, that is, compounds that affect other organisms in the environment and may also evoke a physiological response in humans. Further, these metabolites vary in their concentration depending upon the part of the plant, the season of the year, the chemistry of the soil where they grow—which may explain why herbalists will maintain that there are certain times of year and specific places that are optimal for collecting medicinal plants. These metabolites are often toxic compounds that *in appropriate dosage* are medicines.[9]

Secondary metabolites can be grouped into the following classes of organic compounds, all of which have some action on humans: alkaloids, antibiotics, glycosides, flavonoids, coumarins, tannins, bitter compounds, saponosides, terpenes and essential oils, acids, and mucilages. Thousands of individual compounds constitute these classes.

Alkaloids are organic nitrogenous compounds, thus basic, with specific pharmacological action; they are sometimes very toxic. This toxicity has protected the plants from overgrazing in arid regions such as the American and Mexican West. Plants that grow in hot climates are richer in alkaloids than those from colder regions (Ayensu 1979:119).

Antibiotics are substances that can arrest the growth of microorganisms, such as bacteria, protozoa, and viruses, or destroy them.

Glycosides are organic compounds in which the active constituent, called the aglycone, is combined with one or more glucosides, or monosaccharides. Anthraquinone glycosides are aromatic compounds that on hydrolysis break down to form anthraquinone aglycones, which are cathartics and act on the large intestine. Cyanogenic glycosides form hydrogen cyanide, one of the most toxic compounds found in plants. Phenolic glycosides are weak acids, or phenols.

Flavonoids may act synergistically with vitamin C. They affect the heart and circulatory system and strengthen the capillaries and are generally used as spasmolytics and diuretics (Schauenberg and Paris 1977:108) and possess antipyretic properties.

Coumarins occur in many plants and in more than a hundred different forms. Furocoumarins are formed by the fusion of coumarin with a furan ring.

Tannin is a compound known to cure and preserve leather. The tannins are always found in the bark of the trunk and roots, occasionally in the leaves.

Bitter compounds stimulate the body into reflex action, setting the glands to work and producing their various effects. Plants with the property of bitterness were classified as 'hot' in the ancient Western humoral system (Croke [1607] 1830).

Saponosides are naturally occurring glycosides whose active portions are soluble in water and produce a lather; thus they are used for washing.

Terpenes are the most numerous organic compounds in plants. Essential oils are aromatic; in fact, they are responsible for the fragrance of flowers. They are classified as volatile compounds, monoterpenes, or sesquiterpenes. Sesquiterpene lactones are allergenic chemicals that produce important antiinflammatory effects. They are toxic but also have therapeutic uses. Plants containing essential oils are used mainly as culinary spices today, and many of these are also believed to assist digestion. Others are used as emmenagogues.

Citric and tartaric acids increase the flow of saliva, cleanse the mouth, and reduce the number of bacteria that cause infections or dental caries (Schauenberg and Paris 1977:260).

Mucilages (polysaccharides that occur in medicinal plants) are pectins and gums. In small doses these compounds slow the peristaltic action of the gut, counteracting diarrhea, while in large quantities they can become laxative. When the mucilaginous material passes through the alimentary canal, it coats the organs over which it flows and covers the mucosa with a viscous film; this reduces irritation, slows down the entry of chemicals, and reduces sensitivity to acids and bitters. Mucilages are also good for wounds, being haemostatic and containing antibiotic substances (Schauenberg and Paris 1977:266–67).

It is difficult to assign a specific pharmacological activity of an isolated secondary metabolite to the reputed healing activity of a native plant. There are a few obvious cases where the native plant use unambiguously parallels the activity of the isolated compound: *Salix* bark, salicylic acid; *Quercus* bark, tannin; *Allium sativum*, allicin (see appendix B). In most cases, however, one cannot predict that a metabolite (such as an alkaloid) will explain the reputed health effects of a native plant, although phytochemists endeavor to do this in the literature.[10] In biomedical pharmacy, one active ingredient is emphasized, whereas chemists of herbology say that it is often the interaction of many compounds in each plant prescription that account for the action of the medicinal herbs. Further, these compounds may potentiate the beneficial actions or mitigate the toxicity of other phytochemicals. In describing the phytochemistry of a medicinal plant, most pharmacologists feature only a few of the hundreds of compounds that have been identified and make no provision for possible interactions among them. While it is tempting to use a biomedical rationale to suggest why a particular plant part has been used to treat an illness condition, it is not always appropriate. Finally, the presence of different compounds in a particular plant may explain why entirely different conditions are treated with that plant. For example, agave is used for menstrual pain, wound healing, and colds—each condition perhaps alleviated by a different compound.

Pharmacologically active phytochemicals found in plants of the American and Mexican West are very briefly defined in appendix B. Moreover, presence of the compounds mentioned above will be noted in the individual plant discussions in part 2. Yet these biomedical concepts may fail to address some of the reasons for a given medicinal plant's perceived (if not actual) efficacy. Lay theory of the medicinal actions of plants, or former theory now superseded by biomedical science, may differ from modern biomedical theory. For example, for thousands of years a humoral theory was official, in which plant properties included 'hot,' 'cold,' 'moist,' and 'dry,' and their actions were to counteract the opposite condition in the body (see chapter 3). In traditional herbal medicine, practitioners and patients may have still other cultural reasons for choosing to treat or be treated using plants.

All over the world, people use plants for medicine. Some have no choice: medicines synthesized with biomedical technology are expensive, completely out of reach for many in Third World countries and people who are poor, even in industrialized nations. In addition, some people prefer natural medicines out of conviction or for cultural or emotional reasons. When pharmaceutical preparations fail to cure, chronically or fatally ill people look for alternatives; some find relief, but others may be harmed. Phytochemists are only now beginning to learn the biological activity of some of the compounds in plant remedies—we may be on the threshold of discovering how plants treat illness. Meanwhile, health-care providers can develop respectful knowledge of the choices of their patients.

3 ▪ Illnesses Treated with Plants

Plants have been used as medicines for every kind of problem, from aches to wounds, from abortion to venereal disease. Many health problems have names that are readily recognizable across cultures. In the American and Mexican West, however, some of these illnesses have old-fashioned names that were once the established terms for specific diseases but which are seen today simply as symptoms by biomedicine. ("Fever" is one example.) Some names for diseases carry over from the eighteenth century. Still other names come from the anatomical part believed to be the site of the disease. Folk illnesses, called "Mexican diseases" by Mexican Americans because "American doctors don't believe in them," have other seemingly obsolete names.

These ailments, as translated by the ethnographers or botanists, include external conditions visible on the skin (infections, wounds, animal stings), illnesses characterized by internal symptoms (including conditions of the gastrointestinal, urinary, respiratory, nervous, and cardiovascular systems), diabetes, fever, arthritis, ear and eye conditions, and venereal diseases. Moreover, there are many ways of preparing a plant for curative purposes, many different conditions that may be treated by the same plant, and many folk therapeutic theories underlying the medicinal uses of plants. Thus, the lay vocabularies reflect a complex set of cultural practices and should be examined closely for clues to the symptoms, illnesses, and treatment regimens of the peoples of the American and Mexican West.

"Mexican Diseases"

Some illness labels are old-fashioned Spanish terms in lay vocabulary that do not have exact counterparts in biomedicine or exact translations into English.[1] Labels for these symptom categories come from sixteenth-century official and lay medical theory of Europe and from pre-Conquest Aztec theory. One such illness is *empacho,* a condition of swollen belly wherein undigested food matter is stuck, causing various symptoms from a gastrointestinal infection (see Weller et al. 1993). Another folk disease of the belly is *pujos,* which translates as "straining" or "grunting"; the infant with this diagnosis is believed to become sick from contact with a menstruating woman or with persons who recently had sexual intercourse. *Pasmo,* believed to be stimulated by rapid chilling of the body after it has been overheated by activity or by bleeding (often following surgery or childbirth), is diagnosed by swellings and skin eruptions, not muscular contractions as the English cognate *spasm* might suggest. *Mollera caída* (fallen fontanelle) would be noted in biomedicine as a symptom of dehydration in an infant, but in lay understanding, it is in itself regarded as a disease (Kay 1993).

To further complicate diagnosis, not all lay illness terms are recognized today by Mexican Americans. One name that does not appear at all after eighteenth-century documents in Spanish is *tabardillo.* The description of tabardillo in old manuscripts as a "pestilent fever" with a rash of small red spots suggests that the condition was typhus; perhaps that diagnosis has replaced the older concept. Another obsolete label is *garrotillo* (coming from the form of capital punishment in Spain at the time, strangulation), which meant a throat swollen by inflammation; the term has since been discarded and replaced by names that are more specific for diseases of the throat.

Eighteenth-century diseases whose labels have changed based on new understanding of etiology include *gota coral* for epilepsy (literally, "gout of the heart," which organ in ancient medicine was believed to be the site of the disturbance). *Dicipela* (a variant of *ericipela,* meaning skin disease) derives from the biomedical term *erysipelas,* which is now restricted to a skin condition caused by a Type A streptococcal infection.

Ethnophysiology

In the ancient humoral medicine of the Old World, almost all conditions could be ascribed to an imbalance of the internal milieu. Each person had a complexion that had in balance the humors blood ('hot' and 'moist'), phlegm ('cold' and 'moist'), yellow bile ('hot' and 'dry'), and black bile or melancholy ('cold' and 'dry'). Illness resulted in an imbalance and would be treated with a plant that restored the balance through the opposite property. Thus, if a person suffered from a disease labeled 'cold' in humoral theory, a plant considered metaphorically 'hot' was given. Conversely, a 'hot' disease was treated by a plant remedy categorized as 'cold.' If a stomachache was caused by 'cold,' it required a 'hot' plant such as *chuchupate* (*Ligusticum,* lovage), but a diarrhea that was

FIGURE 5 *Malva parviflora.* a. Normal flower. b. Flower unopened. c. Fruit. d. Carpel with seed enclosed.

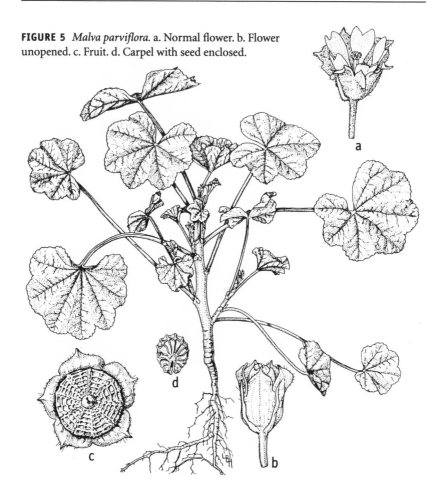

'hot' required a 'cold' plant such as *malva* (*Malva,* mallow). To some extent, 'hot' conditions were those which would be called acute in biomedicine, while 'cold' conditions were likely to be chronic.

Anthropologists do not agree on whether the 'hot' and 'cold' theory of disease attributed to Mexicans and Mexican Americans was indigenous or introduced. Certainly some type of 'hot' and 'cold' theory was symbolized in Aztec illness, as inferred from the work of Father Bernardo Sahagún ([1793] 1956): many herbs were recommended to moderate "internal heat" (Sahagún 1982, 11:670–89) or to treat 'cold' diseases.[2] It seems likely to me that the European and Aztec theories were often combined in the American and Mexican West. Some kind of humoral theory can be found in many cultures, spread sometimes by diffusion of a theory, or invented independently (see Anderson 1988:232–36). The Western 'hot' and 'cold' theory has been found in ancient Greece, Persia, India, China, and the similarity of these ancient ideas to lay

Mexican theory is close. I believe one reason humoral theories are widespread is that they tap universals in healing that are founded on a common pathophysiology of humans reacting to thermal heat and cold (Kay and Yoder 1987). Our perceptions, interpretations, and responses to illness are, then, all based on our particular culture (Kay 1993).

Plant Preparation

Ancient herbals and cookbooks give directions for using plants for healing, including which part of the plant is employed, how to prepare it, and how to administer the medicine. Some give details about harvesting the plant, such as the time of day or year. Ethnobotanies of the American and Mexican West often but not always present such information. Most frequently in ethnobotanies of the region we are told which part of the plant to use.[3] The pharmaceutically active ingredients may be concentrated differently in the root, stalk, branches, leaves, or flowers of the plant; thus it is important for the healer to know which specific parts to select for concocting the plant into medicine. In general, roots are dug before flowering in early spring; biennial and perennial herbs are gathered in late fall when plants are storing nutrients for winter and have stopped growing; bark is gathered in winter or spring when trees can be easily stripped; leaves are generally gathered before blooming and dried in the shade away from dampness; flowers are gathered when they first open; fruits are collected when they are mature.[4]

Different parts of the plant may be used for specific purposes. For example, Aztecs ingested the avocado fruit as an aphrodisiac and the oil for dysentery, or instilled the oil as an enema for constipation. The bark of cottonwood was made into a drink to aid labor, the leaves decocted into a wash for sores, the branches to refresh a sick room or as splints for fractured bones.

Sometimes a plant part is used without modification, such as when leaves are applied directly to a lesion or body part. For example, the Mayo put the leaves of *cacachila* (*Karwinskia humboldtiana*, coffeeberry) to the temples of an aching head and hold them in place by a rag or with grease.

Most often, however, something is done to the plant to prepare it as a medicine. Leaves or flowers may be boiled in water to make a drink, a lotion, or a wash for a skin ailment—for example, *golondrina* (*Euphorbia*, spurge) is cooked and the water used for treating a rash. Juicy plants are roasted where water is scarce. The plant part such as flowers of *manzanilla* (*Matricaria*, chamomile) may be put in a bag before steeping, then removed. The leaves of verbena are crushed to make a poultice. Roots may be washed, then chewed and swallowed (or only the saliva is swallowed, as with marigold). Oil may be expressed from fruit (for example, to make olive oil) or from seeds (as for cottonseed oil). The seeds may be soaked: Mexican Americans steep avocado seeds in alcohol and mix in marijuana to make a liniment to rub on a painful place. For administration through vapor, plant parts may be placed in a steam kettle, over a fire, or on heated stones over which the patient sits: the bitter-tasting brews of creosote *(Larrea)* and desert broom *(Baccharis)* branches are used in these ways. Leaves of tobacco *(Nicotiana)* or mullein *(Verbascum)* are smoked for their imputed effects on asthma and other chest conditions.

The ancients boiled plant parts in wine; today most plants are boiled in water. Solvents for the plant once included urine (specified as to male or female infant) and mother's milk (again specified as to the sex of her infant). The plant may first have been dried, hung with the flower heads down, later to be placed on a flat stone and powdered. Various forms of grease (for example, chicken fat) may be the vehicle for administering the medicine. Homeopaths and natural plant pharmacists make tinctures (alcoholic solutions of the plant part) or put the herb product in capsules or pills. Loose dry parts of one plant or packages of herbal combinations may be purchased in health food stores and in grocery stores, both in the United States and in Mexico. These various means of preparation help to further complicate the process of understanding traditional healing with medicinal plants in the American and Mexican West today.

Conditions Treated with Plants
External Conditions

External conditions may be labeled as boils, bruises, burns, canker sores, fractures, hemorrhoids, herpes, poisonous bites and stings, sores, ulcers, and wounds. Plants have long been used to heal these visible problems as well as the pain they caused. Most conditions were treated by applying the herbal substance directly. For example, the Aztecs used the juice of *sábila (Aloe)*, which had been introduced by the Spanish, for sores and also for erysipelas. *Aloe* is employed by biomedical practitioners today to heal burns, including burns caused by x-ray treatment. The sap of *cacalosúchil* (*Plumeria,* frangipani) and mesquite *(Prosopis)* are also used by many peoples for burns.

Skin diseases were usually considered to be metaphorically 'hot' and thus required a 'cold' remedy. Thus, herpes (which the ancients, modern biomedical practitioners, and a Tucson *curandero* found to be associated with fever) is treated with 'cold' poultices, or *té de nogal negro* (black walnut tea).

The most popular herb for treating bruises, cuts, and ulcers is *yerba del manzo (Anemopsis californica,* lizard tail). Commonly the root is boiled to make a wash for sores, bruises, and lacerations. For abscesses, tumors, inflammations, lacerations, hemorrhoids, and the pain that is caused by tissue swelling or accumulation of pus, *toloache (Datura,* jimsonweed) was often applied as a poultice. The sap or pulverized bark of the root of *Jatropha* species is also widely used for mucous tissue ulcers, including throat, gums, genitals, and hemorrhoids. Similarly, the caustic latex of *golondrina* (*Euphorbia*, spurge) is used for wounds, warts, ulcers, hemorrhoids, swellings, and snake bites.

For cuts and wounds, various Cactaceae have been employed. *Hecho* (*Pachycereus*, organ pipe cactus) was particularly favored in the eighteenth century and also today to check bleeding, especially following a tooth extraction: a wash is made from cooking pieces of the trunk in water. *Nopal* (*Opuntia*, prickly pear) pads were used to heal various kinds of skin breaks. An exotic-sounding but not unusual medication for cracks in the skin was made by frying and mashing a piece of the prickly pear pad, then applying it to the fissures or cracks in skin that had been washed first with a decoction of *malva* (*Malva*, mallow), *golondrina* (*Euphorbia*, spurge), or urine—one's own, a baby boy's, or a dog's. (It is interesting that specifically human male urine and dog urine contain epithelial growth stimulating hormone, and creams containing urea are recommended in biomedicine for severely dry and fissured skin.)

Poisonous insect stings and animal bites have long been treated with plant remedies. Some, such as *Baccharis* (desert broom) teas, are applied directly to the sting; others are taken orally, such as *confituria* (*Lantana*, lantana). Most are purgatives. The underlying folk theory was that the poison would be eliminated from the body through the feces. Various plants used as antivenoms are called snakeroot in English and *yerba de la vibora* in Spanish; several others are called *contrayerba*, "antidote."

In the eighteenth century there were many uses for plants believed to be antidotes for poisoned arrows or for snake and animal poisons, including the sap of *jicamilla de julimes* (*Jatropha*, limberbush) and *huisache* (*Acacia*, acacia). Some plants considered to be antivenoms are themselves highly toxic, such as *tabachín* (*Caesalpinia*, Mexican bird-of-paradise), used by Pima Bajo and Warijio. Some plants are specific to the sting: *torote* (*Bursera*) for the stingray and *huirote* (*Sarcostema*) for the black widow spider for the Seri, who calm the pain with a tea of *torote prieto* (*Bursera laxiflora*, elephant tree). In Baja California, the black widow bite is treated with *fresno* (*Ludwigia octovalis*), while scorpion stings and snake bites are treated with a wash of *golondrina* (*Euphorbia*, spurge). Mexican Americans rub the stung area with garlic.

Loose teeth and infected gums have been problems of mankind since earliest times. *Tepeguaje* (*Lysiloma*, featherbush) and *cañaigre* or *yerba colorada* (*Rumex*, dock) are used extensively for these conditions as well as for sore throat. Esteyneffer ([1719] 1978:362–68), who noted that the disease called in Latin *morbus scorbuticus* (scurvy) caused bleeding gums and mouth sores, recommended a mouthwash made with plants such as *acederas* (Spanish) or *xoxocoyolli* (Nahuatl), an *Oxalis*, *llantén* (*Plantago*, plantain), and *mastuerzo*—or *oyvari*—(*Tropeolum*, cress). Lemon juice, he said, was also good. *Mezcal* (fermented agave drink) and *pitahaya agridulce* (organ pipe cactus fruit) were recommended by other eighteenth-century writers. It is interesting that all of these plants have a high vitamin C content.

Internal Conditions

Internal conditions are also treated with plant medicines. Indeed, the greatest number of entries in the data bank (332) is for plant remedies given for gastro-intestinal disturbance, agreeing with the findings for Mexico generally (Lozoya, Velázquez, and Flores 1988). Many of these remedies are laxatives; thus people suffering from colic, diarrhea, dysentery, and stomachache may be given a purge to be cured. It may appear puzzling that many folk remedies for diarrhea are purgatives; the theory underlying this apparent contradiction is that the body contains something that should be eliminated and that the bowel tries ineffectively to do this by diarrhea.

Empacho is a condition of swollen belly wherein undigested food matter is stuck somewhere in the gastrointestinal track (Weller et al. 1993). There is an "internal fever" that cannot be observed but betrays its presence by great thirst and abdominal swelling, which comes from drinking a great amount of water to quench the thirst. Children, who are prone to swallow chewing gum, most commonly suffer this illness, but its incidence is not restricted to any age group. Although the English word *impaction* would seem to be a logical translation, the cognate does not include the idea of infection, a criterion for diagnosis in the American and Mexican West. *Añil* (*Indigo suffruticosa,* indigo) is supposedly used to treat the condition, but bluing balls, which are simply dye for whitening laundry, are more commonly employed.

The urinary system is also believed to be in need of cleansing to treat frequency, stoppage, and stone. Plant remedies to treat urinary problems generally are prepared as a tea, although some, like agave sap, are used both as a tea and as a poultice applied to the back. Many have diuretic action. Mexican Americans know the tea that is made from *pelos de maíz* (*Zea mays,* corn silk), which they are given as a diuretic. *Alfilerillo* (*Erodium cicutarium,* filaree), an invader of gardens in the Southwest, is also used as a diuretic by northwestern Mexicans and by Mexican Americans for urinary disorders. *Palo cuate* (*Eysenhardtia,* kidneywood), a plant remedy used for kidney problems, was also used for "dropsy," an old-fashioned word for ascites or edema. *Cola de caballo* (*Equisetum laevigatum,* horsetail) is known worldwide for urinary or kidney complaints.

In the American and Mexican West, the most common remedies for heart problems are the Aztecs' *zapote blanco (Casimiroa edulis)* and the Arabs' *te de azahar* (tea from various citrus blossoms, especially orange blossoms), brought to Mexico by the Spaniards: both function as sedatives. Heart symptoms are interpreted in various ways, given different names, and treated with many plant remedies. For example, the Tepehuan use a tea of *Eryngium* (button snakeroot) for heart palpitations, *gordolobo (Gnaphalium)* for heart pain, and *lecheguilla (Senecio)* as a heart stimulant. Similarly, in Baja California Sur tea made from shoots of *Haematoxylon brasiletto* is used for circulation and heart trouble, while "heartache" is treated with *chisme (Pilea microphylla)*. Each ethnic group has plant treatments for cardiovascular symptoms, including conditions such as "heart pain" and "heart warm" (Lozoya 1980:93). Many of these plant remedies are Cactaceae and have demonstrated cardiovascular activity in dogs. Mexican Americans treat elevated blood pressure with some of the same remedies or with other plants such as *ajo* (garlic), *perejil* (parsley), and *sauco* (elder blossom).

American Indians and Mexicans have a high incidence of diabetes as well as many traditional plant remedies to treat the disease. Although all oral hypoglycemics prescribed by biomedicine are derived from plants, none has come from the plants of the American and Mexican West. Many of the popular plants for diabetes do indeed contain compounds with hypoglycemic action. Mexican Americans in Colorado use *altamisa (Artemisia ludoviciana,* western mugwort) or *chamiso (Artemisia tridentata,* sagebrush) for diabetes and consume *nopal redondo (Opuntia,* prickly pear); in New Mexico and Arizona, they also use the roots of *matarique (Psacalium decompositum,* Indian-plantain) or *mora blanca (Morus alba,* white mulberry) and a drink made from the pads of *nopal (Opuntia,* prickly pear) and aloe. Diabetic skin infections are healed with aloe or *maguey.* Newly popular[5] is *guareke (Ibervillea sonorae,* coyote melon), sold fresh or in capsules. The most commonly employed herb is *copalquín (Hintonia).*

Fever is a condition that formerly was differentiated by type. For example, the old-fashioned term *ague* referred to chills and fever, usually malaria. Other fevers were called quartan, tertian, intermittent, and putrefying. The most frequently occurring treatments across cultures were *copalquín (Hintonia)* for malaria or ague and *malva (Malva,* mallow), *sauce (Salix,* willow), and the flowers of *sauco (Sambucus,* elder) for other fevers. All the plants used by the Aztecs for fever appeared in the phytotherapy of some of the other groups: the Aztecs believed that fevers required treatment using an emetic, a purgative, a diuretic, or a diaphoretic (Ortiz de Montellano 1990:220), properties of many plants used for fever.

Arthritic conditions, including *gota* (gout, from Latin word for drop, referring to a "defluxation" of humors) and *reuma* (rheumatism, from the Latin word for flowing), were believed to be caused by fluid collecting in the painful part, "phlegm" in the old humoral system. Mexican Americans treat arthritis with alfalfa, crownbeard *(Verbesina encelioides),* horsetail *(Equisetum),* and burrobrush *(Hymenoclea),* and ease joint pain by applying teas made from aloe, *guamis (Larrea,* creosote bush), *bavis (Anemopsis californica,* lizard tail), or a concoction of marijuana, avocado, and alcohol.

FIGURE 6 *Verbesina encelioides*
var. *exauriculata.* a. Achene.

In the nosology common to the American and Mexican West, diseases are
often named by the place where it hurts; for example, headache is a disease en-
tity. More than one hundred plants are prescribed for different kinds and
places for pain, and these are not always interchangeable: one would not use
the same plants for headache and hemorrhoids.

Respiratory illnesses are common in the American and Mexican West, especially in the colder climates. Plants called *gordolobo,* including both the European *gordolobo* (*Verbascum thapsus,* mullein) and the Mexican *gordolobo* (*Gnaphalium,* pearly everlasting) species, are commonly used. Resin from *torote*
(*Bursera,* elephant tree) was also frequently referred to as treatment. Borage is
an ingredient in cough syrup. Teas made from various species of mint, with
the addition of lemon juice and sweetened with honey, are also used. *Eucalypto*
(*Eucalyptus,* eucalyptus) or *gobernadora* (*Larrea,* creosote bush) leaves are prepared for vapor treatment.

Eye conditions have been treated most often by *cardo* (*Argemone*, prickly poppy) or the resin of mesquite (*Prosopis*), although today's Mexican Americans avoid them. *Ruda* (*Ruta graveolens*, rue) is the most common treatment for ear problems: the leaves are prepared as drops to be directly applied to the ear canal. Ear drops made from the flesh of the pincushion cactus (*Mammillaria*) are employed by several peoples of the American and Mexican West. For these the medication is held in place with a cotton ball. Other plants for earache include a *Baccharis* (desert broom), *alcanfor* (*Cinnamomum*, camphor), and *albahaca* (*Ocimum basilicum*, basil)—all classified as 'hot' herbs.

For nervous conditions, teas brewed with *toronjil* (*Agastache*, giant hyssop), *azahar* (*Citrus* blossom), and valerian have been especially popular. Also used are *granada china* or *pasiflora* (*Passiflora*, passion flower), *salvia real* (*Buddleia perfoliata*, butterfly bush), *tilia* (*Tilia occidentalis*), *tumbavaqueros* (*Ipomoea*, jalap), and *ruda* (*Ruta graveolens*, rue). These plant remedies are also used for heart disorders because the heart is seen as the source of *nervios*.

Various emotional "states" in the nosology of "Mexican diseases" treated by plants include *tiricia* (separation sorrow), *susto* (comparable to biomedicine's post-traumatic stress disorder), and *nervios* (acute nervousness) or *ataque de nervios* (lay concept of "nervous breakdown"). These conditions are often considered to be culture-bound to Hispanics, but they have counterparts in the lay nosology in other cultures and have been known by other peoples of the American and Mexican West.

Every ethnic group reported plants to treat venereal disease, both gonorrhea and syphilis. Of the twenty-three genera employed throughout the American and Mexican West, most are strong purgatives.

Analysis of the origins of the plants reported as medicines in the data base indicates that the peoples of the American and Mexican West most often have been reported to use indigenous, endemic plants. Plants belonging to the Asteraceae (formerly called Compositae) are cited more than three hundred times; Fabaceae (Leguminosae), more than two hundred times; Euphorbiaceae, eighty-nine times; and Cactaceae, sixty-two times. These plants can withstand a hot, arid environment. By contrast, the more widely studied ethnomedical botanic picture of Native North America is very different, with Lamiaceae (Labiatae), Apiaceae (Umbelliferae), and Rosaceae—plants that need more water and lower temperatures—most common (Moerman 1991).

Surveying the illnesses for which plant medicines are employed in the region, however, shows less difference. As in Native North America, gastrointestinal uses for plants rank first in the American and Mexican West. As Pliny (1938, 7:297) said in the first century A.D.: "The greatest part of man's trouble is caused by the belly. Therefore the tasks of medicine concerned with the belly are very numerous." In frequency, belly problems are followed by skin conditions, including wounds, sores, and lacerations. Ethnic groups that live in montane areas are more concerned with respiratory and rheumatic conditions, while those living in the desert inventory more skin problems.

Despite the persistence of traditional plant remedies among the various culture groups of the American and Mexican West, for certain conditions, plant remedies are no longer feasible. Fourteen of the plants in this study were once made into medicines for eye conditions—and long used—but several of these have been superseded by drugstore remedies. Plants in the genus *Argemone*, for example, are still used for eye conditions by many peoples, but phytochemists consider them dangerous today. The same is true for *Rhynchosia* and *Datura*, which are in addition hallucinogenic.

Other plants that are especially toxic have been replaced by milder remedies. *Plumeria* is a drastic purgative; now that purges are no longer considered advisable, it has probably been replaced by moderate cathartics. *Senecio*, primarily used externally to treat pimples and wounds, had also been taken internally. Its poisonous nature may have been noted, leading to its present abandonment. But might some *Senecio* species be valuable for healing dermatological conditions? There is evidence only that it has been used for many thousands of years (Lietava 1992).

Some plants are too much trouble to prepare as medicine. For example, it is difficult (and also unlawful) to obtain stem material from the giant saguaro cactus *(Carnegiea)* to treat pain, and effective pain relievers are readily available over the counter in pharmacies. Still other plants are no longer considered effective as treatments. For example, *Ruellia* (used for fatigue, colds, and headache) grows mainly in Sonora and Baja California. Never described by Hernández, it has not found its way into the numerous little herbal books that are available to Mexican Americans, who have many over-the-counter remedies for colds and pain that are satisfactory. Another plant, *Simmondsia chinensis* (jojoba), may have proved to be disappointing—not curing cancer, treating sores, or helping childbirth (eighteenth-century expectations)—and is hard to find outside its normal range.

Many people are nevertheless turning to alternative medicine, disenchanted with the power of biomedicine and frightened by its cost; might it not be valuable for them to learn what is beneficial and what is dangerous in helping themselves with plants? And biomedical knowledge could be expanded. For example, there may be plants in the desert waiting to take the place of presently known antibiotics as organisms become resistant, or plants to stimulate the immune system, perhaps even plants to alter metabolism. I hope for a collaboration in the American and Mexican West as elsewhere among phytochemists, health-care providers, and users of plant medicines to take advantage of the knowledge that each has gained thus far, and to propel us toward a greater understanding of the potential that is waiting in common regional plants.

4 ▪ Healing the Illnesses of

Women and Children

Women have always been considered to be especially knowledgeable about herbs. As both practitioners and patients, women have a unique concern with medicinal plants. They are commonly responsible for the health of members of their households, diagnosing illness and prescribing treatment with domestic remedies. They also take care of themselves and their children. Moreover, some of the conditions specific to women have been extended to infants and children as well: through childbirth and lactation, mother and child are inextricably linked. Through the centuries, however, the men who have listed traditional medicines for women's health have been nearly silent about medicines for children, giving directions only to the midwife and wet nurse (Kay 1993). In addition, only those women's conditions concerned with reproduction have been defined as different.[1]

Beliefs about women's physiology have changed little. Worldwide, women are supposed to be fragile (although they have always done much of the hard work).[2] Their supposed delicacy comes from those conditions related to menstruation, childbearing, lactation, and associated "nerves," with treatments for these presumed ailments well documented. However, their therapies are made from plants that are used for men's illnesses as well. Through the ages, continuing in modern clinical medicine, there has been no modification of plant usage for illnesses that are not specific to women. Thus, we do not know whether, for example, plants to treat heart problems or venereal disease are to be used by both sexes and to be prepared in the same way.

If we know little about women's health, we know less about the problems of infants and children. Children are prone to particular illnesses because of their great vulnerability, and they show sensitivity to any medicine, yet knowledge about plants used to treat their special health concerns is sparse. This is a serious lack, for *yerbas* are used in the American and Mexican West for many infant illnesses. Moreover, many plants are more toxic for children than for adults (see Appendix A).

Women

In every culture, some of the plants used to treat illnesses in the general population have been used for women's illnesses as well. Of the one hundred genera featured in this book, for example, fifty-one have female uses.

It might seem that some ethnobotanists, especially men, would have had reservations against inquiring about medicine for women's conditions, or possibly lacked knowledgeable informants. However, the eighteenth-century Jesuits and the military personnel who followed (all men, of course) were very insistent and solicitous in this regard: twenty-five of the plants used to treat women's conditions appeared in eighteenth-century records. In fact, no missionary or secular account of illness in an ethnic group entirely failed to pay attention to women's health concerns. Recently, as well, in an indigenous southern Mexican community men (contrary to expectations) were found to be knowledgeable about herbal remedies for reproductive health (Browner 1991). I too found in my research with Mexican Americans that men knew something about medicinal plants for women.

The categories of medicines used to treat women's conditions include abortifacients (which provoke abortion); contraceptives; so-called emmenagogues or menstrual regulators; fertility enhancers; remedies for childbirth (either to stimulate it or to decrease pain during labor); remedies to deliver the fetus and to expel the placenta or afterbirth; remedies to decrease bleeding during the menstrual period, during delivery, or in the postpartum period; remedies for lactation (to increase or decrease milk flow or to heal problems with the breasts); and finally, medicinal plants to treat the effects of menopause. In many cases, the same plant was used for more than one female problem, for example, those which were believed to proceed from the uterus. In biomedical terms, those medicines that act on the uterus are oxytocics: they abort, speed labor, deliver the infant and afterbirth, shorten postpartum pain, decrease excessive menses, and so forth.

Abortion and Contraception

The ancient, pagan writers had no problem with discussing abortion and contraception. For example, Pliny (1938, 6:147–51) recommended mint, which "is believed to be a hindrance to generation by not allowing the genital fluids to thicken." The Aztecs and the conquering Spanish had viewpoints that differed from each other concerning sexual acts. The Aztecs separated procreation and eroticism: Tlazoteotl was the goddess of fertility, birth, and procreation, while Xochiquetzal was the goddess of erotic love. By contrast, the conquistadores

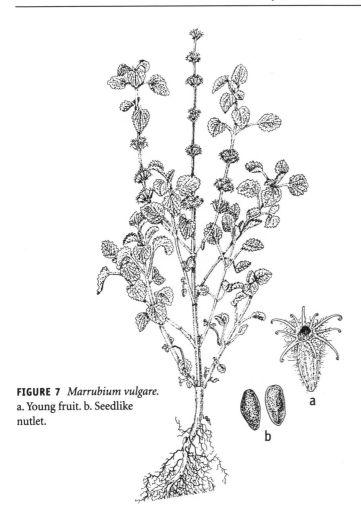

FIGURE 7 *Marrubium vulgare.*
a. Young fruit. b. Seedlike
nutlet.

and the missionaries looked with condemning eyes at "sins against nature," especially the sexual license that they observed in Indian groups (Quezada 1977:224). One such sin was abortion; thus, no plant used to produce an abortion is listed in missionaries' reports for this single purpose.

Traditional pharmacopoeias do contain numerous abortifacients, but the plant remedies are often used for multiple purposes. To induce abortion, for example, Mexicans in Baja California used a decoction of the root, branches, or bark of *gobernadora* (*Larrea,* creosote bush) or of the root or branches of *mantel* (*Mascagnia macroptera,* mascagnia). *Marrubio* (*Marrubium vulgare,* horehound) is (very quietly) recommended in Mexican markets. *Marrubio* has been used since ancient times for its oxytocic activity; Dioscorides noted that it brought on menstruation, expelled the afterbirth or dead child, and helped in labor. Colonial Pima Bajo documents indicate that an *escorzonera (Iostephane heterophylla)* and probably other substances were in fact used to induce abortion. These four plants—*escorzonera, gobernadora, mantel,* and *marrubio*—have also been used as "emmenagogues" (to "bring down" the delayed menstrual period), to stimulate labor, and to aid in childbirth. Such multipurpose herbs as *Ambrosia* (if taken in a sufficiently strong brew) and *Tagetes lucida* (called *yyauhtli* by the Aztecs) have been believed to produce abortion.

In the American and Mexican West, few plants are known specifically as contraceptives, for until recently, underpopulation was the recognized problem. The Indian groups had been scattered as hunters and gatherers; when they were crowded into mission communities, they became prey to introduced diseases, which decimated their numbers. Also, many of the people recording medicinal plants in the eighteenth century were religious missionaries, who apparently did not ask questions about means of avoiding pregnancy except in the confessional (Lavrin 1989). The missionaries instead would ask their charges about "retained menstruation." I also found that it was better to query Mexican and Mexican American women about treatments for "delayed menstruation" and treatments for *la regla,* the menstrual period, than to ask them outright about contraception. With this indirect questioning, respondents are likely to mention folk remedies that are perceived as emmenagogues but that could be considered as contraceptives. In fact, women even now describe the anovulatory pill as a "menstrual regulator."

Nevertheless, a few plants have been used for contraception. The Seri make a tea from the leaves of *batamote* (*Baccharis sarathroides,* desert broom). They make a drink from creosote bush *(Larrea)* by placing on a stick a wad of the lac (an encrustation on the leaves deposited by insects) found on creosote bush, then heating the lac and catching the drops in a container of water. In New Mexico and Arizona rosemary leaves are brewed and the decoction is given as a douche to clean vaginal infection, to regulate menstruation, and also to abort. Mexicans also drink the decoction to cause abortion. Indeed, some refer to rosemary *(romero)* as "the prostitute's herb." A popular abortifacient used by Mexican Americans was a concoction of *Gossypium* (cotton), in which two to four ounces of the bark of the root were boiled in a quart of water for one half-hour and the entire concoction drunk first thing in the morning (Conway and Slocumb 1979).

Emmenagogues

Emmenagogues (as one of the largest categories of women's medicines is known) consist of the plant, mineral, or pulverized animal parts used to treat various disorders of menstruation. An underlying principle in ethnophysiology articulated since the time of the Greek writer of obstetrics Soranus (d. A.D. 138) is that the mature woman, unless she is pregnant, must menstruate.

If she does not menstruate, she might develop poisoning due to accumulated "toxins." Hence many remedies were used to induce bleeding. Although biomedicine now finds that the menstrual "bleeding" is a hormonally determined shedding of the endometrial lining of the uterus, this information was not available to the ancients, the Aztecs, and various groups of the American and Mexican West: all of these peoples were concerned about "delayed," "obstructed," or "insufficient" menstruation.

One plant still used to bring on a late menstrual period is *Perezia*, well known in colonial times. Nentuig ([1764] 1977:61) noted that "a sure remedy is to give them *pipichaguí* [*Perezia*] to drink, . . . even after 3 or 4 months . . . patients almost at the end and at risk of suffocating from the detained blood, with one or two times drinking *cha*, had their blood flow and were healthy. . . . [T]hese poor women have much need of similar remedies because without knowing that it can harm them, [they] enter in the water and bathe at all times, and from this comes the said illness. The sad thing is that they say nothing about their illness and leave themselves to die if the minister when he is called for confession does not ask about what illness they suffer."

Until very recently, it was received knowledge that bathing suppressed menstruation. But as may be inferred from the missionary's statement, pre-Conquest Sonorans did not think that bathing was harmful to the menstruating woman. Only subsequently did they adopt many European gynecological beliefs.[3] In colonial New Spain, the sexual mores of Spain did come to surmount indigenous values. George Foster, in *Culture and Conquest* (1960:122) wondered why the "female culture" of native women was not continued after the conquest of Mexico. It seems likely that the priests actively taught European gynecological practice rooted in the biblical book of *Leviticus* and in the writings of Aristotle, St. Thomas Aquinas, and others.

Regular menstrual difficulties have been thought to result from improper care during the first period experienced by the girl, from wrong diet during the period ('cold' foods should be avoided), or from washing with cold water during menstruation. Indeed, previous generations of Mexican American girls were kept home from school when they were menstruating and were told that they should not bathe, swim, or wash their hair. The potentially chilling effects of immersion in water were considered dangerous.

To induce menstruation, then, folk physiology utilized 'hot' and 'cold' principles. Thus, those plants believed to be "heating" were given in teas, in solutions for douches, in ointments, or to be absorbed by sitting over steam: "heating" herbs included Old World cinnamon, oregano, basil, and garlic. Conversely, because eating metaphorically 'cold' food might cause retention of menses, 'cold' herbs were used to treat vaginal discharge and excessive bleeding. Plants that Hernández identified as preventing miscarriage were of a 'cold' and astringent nature. Esteyneffer said unripe fruit, especially lemon or orange, caused retention of menstruation; a Mexican American woman indeed demonstrated to me how lemon juice squeezed onto meat caused the outside to coagulate, from which I was to infer that the same coagulation occurred in a woman's body.

Another way to induce menstruation was through purging. Some of the medicines taken were mild purges, while others were drastic. *Perezia*, for example, acts as a violent purge; it was used to induce menstruation as well as to treat venereal disease. Menstruation was also believed to be induced by sudorifics (which cause sweating) or diuretics (which cause urination). Thus emmenagogues were believed to aid in menstruation through various actions—heating, purging, sweating, and diuresis.

In the eighteenth century, female physiology was understood to be directed by the condition of the uterus, called in Spanish *la matriz* or *la madre* (literally, "mother"), and the blood. *Mal de madre*, uterine "suffocation," was a serious condition believed to originate from corrupted menstrual blood whose vapors caused various effects in the body, especially swelling and ultimately convulsions (Esteyneffer [1719] 1978:428–34). *Baccharis* and *Haplopappus* were used in teas for this problem.

Fertility Enhancers

To stimulate conception, the putative aphrodisiac *damiana* is the herb of choice for both men and women, to be taken regularly and before intercourse. It is recognized in Baja California as treating impotence, frigidity, sterility, and sexual exhaustion. Mexican Americans say it will correct the cause of infertility, which is *frío en la matriz* ("cold in the uterus") and inflammation.

Frigidity, the opposite of 'heat,' is believed in many cultures to be related to infertility; thus, many fertility enhancers are seen as having the property of 'heat.' *Salvia*, noted for bringing about conception, for example, was categorized as "Hot and Dry in the beginning of the third degree" by Gerard ([1633] 1975:766).

Mexican Americans also recommend purging with *ojasen* (*Cassia*, senna) or *cocolmeca* (*Phaseolus*, tepary bean) to promote conception. A different approach uses the astringent *cuachalalate* (*Juliana adstringens*). New remedies recommended in a Tucson herb store include a combination—called *ovaryl*—of *yerba de pollo* (*Commelina tuberosa*), *damiana* (*Turnera*), and *palo dulce* (*Eysenhardtia*, kidneywood), as well as herbs called *tizana de uva, estafiate de caballo*, and *sensitiva*.

Sometimes sterility is believed to be caused by inflammations of the genitourinary tract. *Tlachichinole* (*Kohleria deppeana*, tree gloxinia) in a decoction is used by Mexican women in a vaginal douche for inflammation of the ovaries or leucorrhea. The decoction is also drunk as a tea.

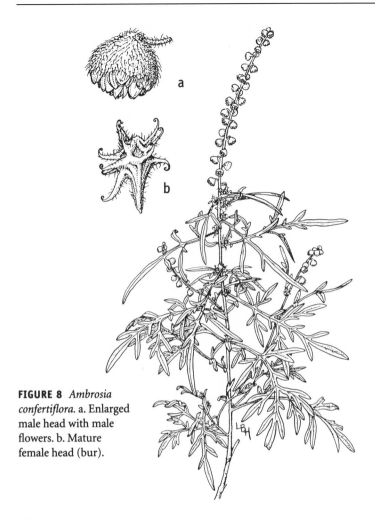

FIGURE 8 *Ambrosia confertiflora.* a. Enlarged male head with male flowers. b. Mature female head (bur).

Childbirth and the Postpartum Conditions

In the data base, there are fifty-five entries of forty-four different genera of plants that aid women in childbirth. Many common herbs have been used to treat the various women's conditions associated with childbirth and its after-effects. *Ambrosia,* for example, the most frequently used plant for women's conditions throughout the American and Mexican West, is given by the Tepehuan and the Mountain Pima as a tea to women who are experiencing difficulty in parturition. Mexicans of Baja California also use a decoction of the root to fortify the uterus after childbirth, and as an abortive, while the Mayo (Kay and O'Connor 1988) cook the leaves in water to make a tea used to clean out the uterus after giving birth.

Anemopsis, also used for a number of women's conditions, is prepared by the Paipai of Santa Catarina in a wash to treat childbirth wounds or by the

Mayo as a douche to treat excessive bleeding after childbirth. Fried in tallow or reduced to a powder, it was applied directly to tears caused in childbirth, according to Esteyneffer, while Nentuig found *golondrina* (*Euphorbia*, spurge), similarly prepared, to be good for healing the fresh wounds of women recently delivered.

Worldwide, the all-purpose agent for women's health problems seems to be *Artemisia*, a genus of plants perhaps named for Artemis, the Greek goddess of women. *Artemisia*, as noted by Pliny in the first century A.D., "doth properly cure womens diseases" (Gerard [1633] 1975:1104). In the twentieth century, Claude Lévi-Strauss (1966:46) found that "in North America, as in the Ancient World, artemisia is a plant with a connotation of feminine, lunar, nocturnal, mainly used in the treatment of dysmenorrhoea and difficult child births." After childbirth, the leaves of *Artemisia tridentata* are brewed with baking soda and salt in a tea taken twice daily for six weeks to decrease swelling.

Other common plants used for childbirth problems are *Larrea* (creosote bush), *manzanilla* (*Matricaria*, chamomile), and *Salvia* (sage). The name *Matricaria* encodes the Latin term *matrix*, "womb." *Matricaria*, said Gerard ([1633] 1975:652), "is a great remedy against the diseases of the matrix; it procureth women's sickness with speed; it bringeth forth the afterbirth and the dead child, whether it be drunk in a decoction or boiled in a bath and the woman sit over it; or the herbs sodden and applied to the privie part, in manner of cataplasme or poultice." *Salvia* was called the holy herb by the ancients because it helped women's childbearing. Indeed, Gerard ([1633] 1975:766) quoted Agrippa and Aetius as saying, "If the woman about the fourth day of her going abroad after her childing [40 days after the birth], shall drink nine ounces of the juice of Sage with a little salt, and then use the companie of her husband, she shall without doubt conceive and bring forth store of children, which are the blessing of God." The Ópata also used a *Salvia* to promote menstruation and facilitate childbirth. It was believed good for women to drink it in tea for a few days after childbirth. The Mountain Pima drink this tea before and after childbirth.

Indigenous plants are also used. For example, the Tarahumara make a decoction of the flowers, stem, and leaves of *Hyptis emoryi*, which they call *chía*, for women in childbirth. The Seri give a tea of *mantel rojo* (*Mascagnia*) to help women gain strength after childbirth. The Seri use the tea of *yerba del venado* (*Porophyllum gracile*, odora) during difficult delivery; the baby is said to be born quickly, disliking the plant's bad odor. The natives of lower California were reported to give five or six jojoba nuts in broth or wine to facilitate women in giving birth, when the urge to push comes (Clavijero [1786] 1937).

Another plant used for women's problems in the American and Mexican West but not elsewhere is *inmortal* (*Asclepias asperula*, spider milkweed): a little of the finely ground root is given in cold water or rubbed on the abdomen for labor pains, or a hot decoction of the powder is given to facilitate expulsion of the placenta.

Other Women's Conditions

Various other conditions were thought to afflict women and to require treatment in the American and Mexican West. These included excessive bleeding, vaginal discharge, menopause, and difficulties with lactation. Again, many of the plants used to treat women's conditions were also used for other illnesses. In addition, plants used for one women's condition were often applied to others.

Chenopodium (wormseed), for example, is used widely for women, including action as an emmenagogue, an abortifacient, and a galactogogue as well as to treat postpartum pain. For excessive menses a tea of *Gutierrezia* (snakeweed) species is drunk by Mexican Americans, who use it for various female complaints. They brew a tea from the branches, which they take four times daily until the pain is gone. It is said to "warm the ovary." The Pima crush and boil the roots of *Ambrosia* and administer the decoction for women's pains and menstrual hemorrhage. Mexican American women make a douche from *Ambrosia* leaves "to clean everything out" after menstruation and also take a tea made of the leaves for menopausal symptoms. Another popular vaginal douche is made from the boiled shells of *nogal* (walnut) or the bark of *nogal silvestre* (pecan).

Artemisia, which has more than fifty entries in the Ethnoagents in Women's Health Care data bank, is used in nineteen categories, including delayed, absent, or irregular menses; insufficient or heavy flow; menstrual pain or other unspecified menstrual problems; leucorrhea (white discharge), vaginitis, pruritis (itching); or other unspecified female problems. Esteyneffer ([1719] 1978: 432) recommended *Artemisia* species for retained menses, difficult childbirth, retained placenta, *mal de madre*, and indurated breasts. New Mexican Spanish-Americans use *Artemisia franserioides* for menstrual pain and the menopausal hot flash. They use *Artemisia tridentata* (big sagebrush) in an external poultice over the umbilicus and small of the back for menstrual cramps and to heal and prevent infection after miscarriage (Conway and Slocumb 1979:250).

Breast problems are given little attention in the various documents, although Esteyneffer recommended *manzanilla* (chamomile) to treat hardened breasts as well as retained menses and to bring about placental expulsion. Mexican Americans in New Mexico and Arizona rely on *albahaca* (*Ocimum*, basil) to decrease their milk supply, even as Dioscorides had recommended in the first century A.D. Many ancient ideas of lactation are followed by Mexican Americans. For example, Esteyneffer ([1719] 1978:445–50) recommended eating rue, cumin, the seed of cannabis, coriander, willow, or plantago to decrease mother's milk, which he believed necessary to prevent inflammation of the breast. It was believed that the excess milk coagulated in the breast (called "caked breast"). Applications of rose oil or *chichiquelite* (*Solanum nigrum*, nightshade) were among his preferred cures if inflammation developed. *Manteca de cacao* (cocoa butter) was good for fissures on the nipple and is still used.

Cancer of the breast, *zaratán*, generally had no cure—although an ointment of green frogs, as well as other animal-derived medicaments might help, Esteyneffer said, in its beginning stages. Jojoba was good for all kinds of cancer, and mashed seed of *adormidera* (poppy) mitigated the great pain (Esteyneffer [1719] 1978:596–604).

Infants and Children

Little information has been collected on treatments for children, a tradition of neglect noted as having begun as early as 1659 for European herbals (Radbill 1974). For example, Aztec medicines for children were virtually ignored by Sahagún and Hernández. Those writing today of medicinal plants in the American and Mexican West for children are usually women authors or co-authors (examples are L.M.S. Curtin, Margarita Kay, Edelmira Linares, Mary Beck Moser, and Marianne Yoder). More is recorded about medicinal plants for babies because infant care information usually follows discussions of childbirth in herbal books and ethnographic records.

Infants

Across the cultures of the American and Mexican West, 'cold' was seen as the most serious threat to the health of the infants. However, many other concerns—such as preventing infection of the umbilical cord stump and eyes of the newborn, relieving colic, promoting sleep, relieving the itching of diaper rash, treating diarrhea, and raising the fontanelle—also received attention.

When the Tepehuan newborn's umbilical cord is cut, the root of *Chromolepsis heterophylla* is applied as a poultice. The Pima and their Maricopa neighbors pack the powdered root of *chacate* (*Krameria grayi*, ratany) on the cord stump after the newborn's umbilical cord is cut; it is believed to give "absolute prevention" from infection of the navel. Mesquite gum is pounded into powder, mixed with very fine sand, tasted, and (if found not too bitter) sprinkled on the navel to prevent infection. The Seri apply the leaves of *torote* (*Bursera microphylla*) to help dry the cord stump.

In the 1940s, the newly delivered Spanish-American infant was bathed with *amole* (*Agave*) in New Mexico. Lemon juice was dropped into the eyes to prevent newborn eye infection. After the navel cord stump fell off, *punche* (*Nicotiana*, tobacco) or *romero* (*Rosmarinus officinalis*, rosemary) was put on the stump (van der Eerden 1948:19).

Today, many Mexican Americans give tea made of *saúco* (*Sambucus mexicana*, elder flower), *yerba buena* (*Mentha spicata*, spearmint), *manzanilla* (*Matricaria recutita*, chamomile), or *albahaca* (*Ocimum basilicum*, basil) to the infant for relief of colic, or *anís de estrella* (*Illicium verum*, star anise), which is also given to promote sleep. If the breast-feeding mother (and breast-feeding is not common today among Mexican Americans) eats food that is gas producing, she gives her infant a tea of *epazote* (*Chenopodium ambrosioides*, wormseed) leaves.

A tea of rose petals treats fever. It also is used as a gentle purge for children. Honey may be added, if it is for an infant (this is a dangerous practice because of the danger of tetanus contaminating the honey). For a *tos de calor* (hot cough, that is, acute) the tea is drunk. To relieve itching of diaper rash, dry rose petals are ground to a powder and applied to the rash. The *chinqual* (rash) may be cured by giving rose-flower tea in the infant's bottle. Rose-flower tea is also given for baby colic and *torzones de tripas* (intestinal cramps) (Bye and Linares 1986:298).

A pacifier is made by wrapping the leaves of *poleo* (*Hedeoma pulegioides*, American pennyroyal) with sugar in a small cloth and dipping it in warm water.

The Aztecs laid the mashed leaves of *cañafistula* (*Cassia occidentalis*, senna) on the stomach of infants who vomited their milk (Hernández 1960:130). Baja California infants receive a decoction of oregano *(Lippia palmeri)* for colic. The Aztecs treated infant diarrhea by applying mashed *zapote (Casimiroa edulis)* to the breast of the wet nurse (Hernández 1959, 2:92). Mexican Americans treat infant diarrhea with a thin *atole* of 2 tablespoons of corn starch with cinnamon and water; rice water is given instead of milk.

The fontanelle, located at the top of the baby's head, is a concern in every society (Kay 1993). It was of great concern to the Aztecs because the *tonalli*, the animistic force that supplied heat to the body, might be damaged or escape through the fontanelle (López Austin 1980; Ortiz de Montellano 1990). Compared to the rest of the skull, fontanelles feel soft to the touch, hence the term *soft spot* or *mollera*, from the Latin *mollis*, soft. The folk nosology of Mexicans and Mexican Americans includes a condition called *mollera caída* (fallen fontanelle), which in biomedicine is known to be caused by dehydration, usually from gastroenteritis or other infection. Hernández listed the ancient treatment with rose oil (cooling for a 'hot' condition) as an ingredient in the treatment for *siriasis*, as the condition was known in ancient Greece and Renaissance Europe. Sahagún noted that in the Aztec treatment, tomato was pressed on the palate. In eighteenth-century Sonora, *copal* (*Bursera jorullensis*, incense) was mixed with egg white and applied as a plaster to raise the fontanelle (Esteyneffer [1719] 1978:441). Pima Alto apply *Ruta graveolens* (rue) to the fontanelle of the sick infant after it has been "raised" by pressure on the palate or suction on the head. The Paipai use similar treatments. To harden the fontanelle the Seri place the peeled and ground root of *jaramatraca* (*Wilcoxia striata*, night blooming cereus) mixed with salt or mother's milk on the baby's head.

Children

Children's conditions, as mentioned earlier, received little attention from observers recording herbal medicine in the American and Mexican West. Nevertheless, at least a few children's conditions have been documented as receiving treatment using medicinal plants. Beyond the childhood diseases chicken pox and measles, children have also suffered from fever, diarrhea, and intestinal worms and have required soothing from the pain of illness or from fright, bad dreams, or emotional illness. Still another concern was delayed walking, which was seen as a physical problem.

The Seri cook an indigobush *(Dalea parryi)* in water to make a tea that is to give strength for walking to the youngest child of a pregnant woman, or to a child of walking age who showed no ability to walk. A lotion was made from the interior portions of branches of *brasil* (*Haematoxylon brasiletto,* brazilwood) to treat small Pima Bajo boys who were ill. A Pima Alto child crying out in his sleep was soothed by applying chewed roots of *cachanilla* (*Pulchea wericea,* marsh fleabane) all over his body.

The Seri make a tea from the leaves and stems of *ojasen* (*Cassia covesii,* senna) for chicken pox. For fever, a child is bathed with *orcilla* (*Rocella babingtonia,* lichen) that was ground on a metate and squeezed through a cloth with water. Mexican and Mexican American children are given a tea of *saúco* (*Sambucus mexicana,* elder flower) for chicken pox and also for measles. Mayo children are given a tea of *chíchivo (Artemisia mexicana)* and valerian for stomachaches.

Esteyneffer in 1712 recommended an application of cottonseed oil to an encrusted scalp, and this treatment is still followed today. The Mexican American child with diarrhea or *empacho* (see chapter 3) is given *Artemisia* brewed in a tea made from only two leaves (because the brew is bitter). Another bitter-tasting medicine, a *yerba del pasmo,* is administered by vapor. *Larrea* (creosote bush) is one of nine plants used for worms, most prevalent in children; another is *epazote (Chenopodium).* Fossilized *Chenopodium* pollen was found in archaeological excavations of southwestern caves with evidence of parasitic disease. The plant is still recognized as a vermifuge throughout the world.

Finally, medicinal plants are used to treat emotional illness in children. The Mexican child is cheered by being bathed in water reddened with *brasil (Haematoxylon brasiletto)* if he or she is jealous of the new baby in the family. For *espanto* (fright), Mountain Pima children are given a drink made from *siempreviva (Echeveria simulans,* stonecrop).

Biomedical health-care providers should encourage Mexican American parents to give children teas made from nontoxic plants such as *Nepeta* (catnip), *Matricaria* (chamomile), and *Mentha* (mint) because these teas contribute needed fluid. They should also point out that the tea should not be sweetened with honey because of the danger of tetanus spores in the honey.

Biomedical health-care providers should become knowledgeable about medicinal plants for women's problems because so many of the same ones are used everywhere. As the globe shrank in the past centuries, medicinal use of many plants in specific genera converged throughout the world. We are now learning something about the biological activity of these universally utilized plants. However, the chemistry of plants used by isolated groups such as the Seri, Pima, and Warijio has been studied little. Thus, health-care providers cannot know whether these plants are safe or efficacious.

Health practitioners can, however, become conversant with women's folk ideas of physiology. For example, normal menstruation is perceived by many to be important to health. Thus, there are many remedies to take if the period is "late." The same plant medicines employed for menstrual regulation are known as contraceptives, even abortifacients, by those who accept such action. They are also used to aid childbirth, as oxytocics. But are they safe? When some people choose traditional healing and many others have no alternative, we must learn about their medicines.

2.

▪ Medicinal Plant Listings

Part 2 is organized by plant genus rather than plant species because this has proven to be the most efficient way to handle the plethora of plants, since in most cases certain species in a genus all have the same common name in Spanish and are similarly employed as medicine. This is not true in every instance; exceptions are treated separately.

Key to Plant Listings

▪ **_genus name_** (family name)

COMMON NAME OF GENUS

Species name
common name of species

Plant description

genus name
1. Common medical uses
2. Known activity of the plant constituents
3. Toxicity

▪ **_Capsicum_** (Solanaceae)

CHILI

Capsicum annuum L.
ginnie pepper, Renaissance English; *chilli*, Aztec; *korí*, Tarahumara; *chili*, Spanish; chili, chili pepper, English

Chilis are small herbaceous plants bearing dry, strong-tasting fruits that turn red as they mature. Wild chiltepine fruits are the size and shape of peas.

Capsicum
1. Used against pain.
2. Capsaicin stimulates salivation and sweating, reduces blood coagulants and triglycerides, and strengthens heartbeat.
3. Is high in vitamin C. Can produce dermatitis and irritate mucous membranes.

Each plant presentation is headed by the name of the genus and family in Latin scientific nomenclature, followed by the English common name, if one is known. The different species in the genus that are used as medicines are then listed by Latin name along with the common names in the various languages that attach to each species (if one has been recorded; in many cases, the Spanish name is used by ethnic groups). Where no species identification is available, plants are listed simply under the name of the genus. Under the list of names follows a very brief, general description of the plants in the genus, where applicable including a discussion of the derivations of the names. Information about the medicinal uses of the most common species in each genus is summarized close by: (1) common medical uses today, especially among Mexican Americans but occasionally by other peoples when Mexican American uses are not recorded, (2) known activity of the plant constituents, and (3) toxicity. The remainder of the genus information includes the following:

history of medicinal use by Aztecs, people of the Old World, and peoples of the region during the eighteenth century (see chapter 1)

medicinal uses of the genus in modern times, organized by cultures and language groups (for sources, see bibliographic essay)

phytochemistry of the plants in each genus, noting compounds reported by phytochemists to be active

I used the stock inventories of medicinal herb stores and grocery stores in Tucson in October 1995 to verify the present availability of plants. The plant parts are packaged in packets of one to two ounces and are labeled with the common Spanish name. Occasionally the botanical name is listed on order forms; however, these names may be misspelled, archaic, or incorrect. The source for the packet labels as well as the numerous booklets sold with the packets is usually *Plantas medicinales de Mexico* (1969), written by the eminent Mexican botanist Maximinio Martínez in 1934 and reprinted many times.

■ *Acacia* (Fabaceae)

CASSIE, ACACIA

Acacia constricta Benth.
oeno-rama (good branch), Seri; whitethorn acacia, English

Acacia cornigera (L.) Willd.
huitzmamaxalli, *púas de piedra* (thorns of stone), Aztec

Acacia cymbispina Sprague & Riley
[= *Acacia cochliacantha* S. Wats.]
chirowi, Pima Bajo; *súsigai doadigami* (pain remedy),
   Tepehuan; *kuduri*, Warijio; *guinole*, Baja California;
   vinorama, Spanish

Acacia farnesiana (L.) Willd.
vinorama, Mountain Pima; *mokowí*, Tarahumara; *kuká*,
   Warijio; *kukka*, Mayo; *cuca*, *cuhuca*, Yaqui; *huisache*,
   guisache, *binorama*, *vinorama*, Spanish

Acacia greggii A. Gray
jussi, Ópata; *uña del gato*, Spanish; catclaw acacia, English

Acacia pennatula (Schlecht. & Cham.) Benth.
algarroba, Tepehuan

Acacia spp.
tezo, Baja California Sur; acacia, English

Plants in the genus *Acacia* are widely distributed throughout the world in hot climates. Acacias are among the more
common desert trees in the Sonoran Desert. They occur
mostly along arroyos but also on hillsides and have numerous tiny leaflets, clumps of small yellow flowers, and beanlike fruits. Most species have spines. *Acacia* gets its botanical name from the Greek *akis*, a sharp point, referring to
the thorns. The sixteenth-century names also emphasize
the sharp thorns, including *púa* in Spanish and the *hui*-
prefix in Nahuatl. *Binorama* is a hybrid word; *bino-* is Cáhitan (from *binole*, acacia) and *rama* is Spanish for branch
(Sobarzo 1966:43).

Historic Use. In the Old World, Gerard ([1633] 1975:1330–
31), writing of "Egyptian thorne," quotes Galen as saying
that *Acacia* has both thin and 'hot' parts dispersed in itself;
therefore it is 'dry' in the third degree and 'cold' in the first

Acacia

1. Used for pain, urinary complaints, and upset stomach.

2. Contains emollients and salicylic acid and exhibits antibiotic activity, depending on species.

3. Pollen may be allergenic.

if not washed, but 'cold' in the second degree if washed. It soothed eye, mouth, and skin inflammations and was also used for pain and for heavy menstruation as well as gonorrhea (by which was meant seminal emission).

The Aztecs (Hernández 1959, 3:275) applied the leaves of *A. cornigera* to previously scarified wounds to combat poisons and the bites of venomous serpents. An eighteenth-century missionary (Nentuig [1764] 1977:64) reported that *uña del gato* mashed and beaten well with water, then strained, was used for urinary complaints.

Modern Use. The Pima Bajo apply the clean bark of *A. cymbispina* to the head to relieve a severe headache and boil the flowers and spines to make a drink for alleviating stomach pains. The Warijio and Mayo boil the roots or widely flaring spines for stomach complaints. For diarrhea and smelly vomit, the Mayo cook strips of the bark and place these on the wrists and feet until cured. The Warijio mix the flowers of *A. farnesiana* with grease to rub on bruises or on the forehead for headache. The Yaqui use the roots and twigs of an *Acacia* species, ground and mixed with water and rubbed on the body, to relieve sleeplessness. They take a strip of bark one inch wide and two inches long, beat it to a pulp, and apply it as a poultice for a headache, leaving it on until it makes an offensive odor.

The Tarahumara cook the bark and spines of *A. farnesiana* to make a drink to treat scorpion stings. The Mountain Pima prepare a drink from the bark and flowers for fever. The Baja Californians Sur make a tea from the branches of *A. cymbispina* that they take for urinary complaints, including "kidney ache, cystitis, and urethritis." Leaves of the same species are used by the Tepehuan not only for urinary difficulties but also for chest pains. The Tepehuan take the bark of *A. pennatula* "from the side of the tree upon which the sun rises," making it into a drink to ease the discomfort caused by venereal disease. The Seri mash the seeds and leaves of *A. constricta* to make a drink for diarrhea and upset stomach.

Mexicans of Baja California Sur fry the bark of an *Acacia* species they call *tezo* in oil; this preparation they apply to the ear for earache and also use to treat sinusitis. Although none of the Mexican American informants in Tuc-

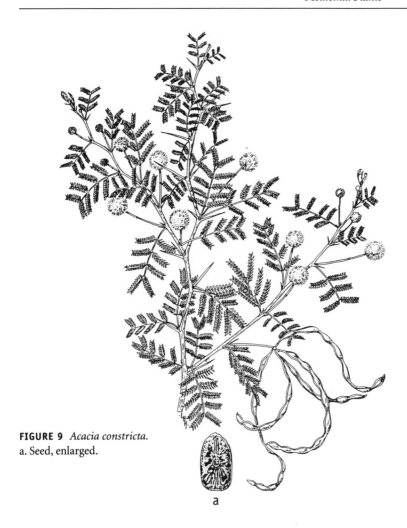

FIGURE 9 *Acacia constricta.*
a. Seed, enlarged.

a

son reported using *Acacia,* several of these species are sold as medicinals in supermarkets.

Phytochemistry. *Acacia* species contain tannins and the aromatic amine tyramine (Duke 1985:7). Numerous compounds have been identified for *A. farnesiana,* including anisaldehyde, benzoic acid, coumarin, cresol, methyleugenol, methyl salicylate, eugenol, and terpineol. *Acacia senegal,* an African species, yields gum arabic, which is used internally in treating inflammation of intestinal mucosa and externally to cover inflamed surfaces. Possibly the gums of other *Acacia* species also have emollient properties. An extract of the branches of *A. cymbispina* was found to have antibiotic activity against *Staphylococcus aureus* and *Bacillus subtilis* (Encarnación and Keer 1991); *Streptococcus pneumoniae* was found to be inhibited by the leaf of a Guatemalan species, *Acacia hindsii* (Cáceres et al. 1991). *Acacia* pollen may be allergenic.

■ *Achillea* (Asteraceae)

YARROW

Achillea lanulosa Nutt.
ha'tsenawe (cold leaf), Zuni; *plumajillo,* Spanish; western yarrow, English

Achillea millefolium L.
tlaquequetzal, pluma de la tierra (feather of the land), Aztec; *hazéíyiltsee'í* (chipmunklike tail), Navajo; yarrow, milfoil, nosebleed, English

Achillea lanulosa is a native plant; *A. millefolium* was introduced from the Old World. The two species differ only by chromosome number. Yarrow is a common weed with many feathery leaves (hence the name *millefolium,* thousand leaves) and a strong, rather distinctive odor. Near the top of the plant are numerous tiny white flowers. It flourishes at elevations above 5,000 feet. This genus was perhaps given its botanical name for Achilles (Pliny 1938, 7:169), who was supposed to have treated his soldiers' wounds with it. The Nahuatl name incorporates *quetzal,* feather.

Achillea

1. Used externally for wound healing.

2. Antiinflammatory and hemostatic; taken internally, can be antipyretic and abortifacient.

3. May cause dermatitis; contains thujone, which is toxic.

Historic Use. Hernández (1959, 3:172) found that to cure a cough, the Aztecs drank *tlaquequetzal,* which purged all humors by vomit. It was recommended for women to alleviate "uterine strangulation" and to hasten childbirth when delivered as a vapor bath, suppository, or drink. It also provoked urination. Applied as a plaster to sores, it cured ulcers of sexual parts, resolved tumors, and treated scabs on the infant's head. Pulverized and applied, or made into a drink, it treated stomach complaints including flatulence and diarrhea and strengthened a stomach weakened by 'cold'. Hernández described *tlaquequetzal* as 'hot' and 'dry' in the third degree, with acrimony.

For the Old World, Gerard ([1633] 1975:1073) made all the same recommendations as Hernández, but his *milfoile* was described as 'cold', not 'hot'. Pliny (1938, 7:377) said that excessive menstruation was checked by an application of *Achillea* or a sitz bath in a decoction of it.

Modern Use. The Zuni said that the leaves of a plant identified as *A. lanulosa* produce a cooler sensation, when applied to the skin, than any other plant (Stevenson 1915:42). The Navajo use yarrow for fever and headaches and for healing sores on people and animals; they say it is especially good for healing saddle sores on horses (Mayes and Lacey 1989:137–38). Mexican Americans make a tea of the herbage of *plumajillo* that anemic persons drink to enrich the blood. For a cold, they combine *plumajillo* with *Mentha, Artemisia arvensis,* and *Artemisia frigida* to make an infusion, a vapor for inhaling, or a preparation for *jumaso*— steam in a hot bath. A tea of the herbage treats *el frío en el estómago,* pain caused by cold air, and also painful and bloody urination.

Members of the genus *Achillea* were used medicinally throughout Alaska for nearly every type of sickness or injury (Fortuine 1988:214). The medicinal use of *Achillea* in various parts of the world is especially intriguing, since Old World peoples used *A. millefolium* and pre-Conquest New World peoples used *A. lanulosa* for similar purposes.

Phytochemistry. A few of the more than 140 constituents of *Achillea* have been isolated (Kelley, Appelt, and Appelt 1988:6–8; Lietava 1992:264). *Achillea* contains azulenes and sesquiterpene lactones, which possess antiinflammatory effects and manifest membrane-stabilizing effects with ensuing inflammatory edema reduction. *Achillea* has a high content of tannic acids, the flavonoids rutin and quercetin, aspirinlike substances, and chamazulene in the volatile oil. A sesquiterpene lactone, thujone, possesses abortifacient properties. It should be noted that sesquiterpene lactones are toxic and tannins are carcinogenic.

■ *Agastache* (Lamiaceae)
GIANT HYSSOP

Agastache mexicana (Kunth) Lint & Epl.
tlalámatl, Aztec; *toronjil,* Baja California Norte, Spanish

Agastache micrantha (A. Gray) Woot. & Standl.
húpachí, té de menta, Tarahumara

Agastache pallida (Lindl.) Cory
toronjil, Tarahumara; giant hyssop, English

Agastache spp.
yerba de la virgin, Mountain Pima

Giant hyssop, the common name for the whole genus, is a member of the mint family having oblong to triangular leaves and a long spike of clumps of small red or pink flowers. *Agastache* receives its name from the Greek *agan* (very much) and *stachys* (a spike), words that refer to the numerous flower spikes (Coombes 1985:21). The common name *toronjil* is also applied to other species in the mint family.

Agastache

1. Used for nausea, vomiting, and nerves.

2. Some species have antispasmodic and antiinflammatory activity.

3. No reports of toxicity from use.

Historic Use. In the *Badianus* (Emmart 1940) it is reported that the Aztecs used drops pressed from the root to cure wounds; Hernández said that the powdered root cured ulcers, mitigated pain and fever, and served to evacuate humors (Linares, Bye, and Flores 1984:72).

Modern Use. Several *Agastache* species are employed in the American and Mexican West. The Tarahumara use one species of *Agastache* for colic and another for cold and cough as a decoction. They place a pinch of the leaves in the nostrils to clear the head. The Mountain Pima use an *Agastache* to make a hot drink for stomach disorders. In Baja California Norte, *toronjil* is recommended for stomach problems: nausea, vomiting, poor digestion, and gastritis. It is also for the nerves, treating hysteria, nervousness, and loss of emotional control. Indeed, it is sold in Mexican herb stores as *anti-nerviosa* and is an ingredient in a "super special" tonic prepared by a Tucson *curandero*.

Phytochemistry. Interested researchers have found no studies specific to *A. mexicana* (see Winkelman 1986:112), but other *Agastache* species contain the aglycones and flavonid glucosides luteolin, apigenin, and acacetin. One species was found to contain methyleugenol. The toxins anisaldehyde and pulegone have also been reported (Duke 1985:547) in *Agastache* species.

■ *Agave* (Agavaceae)
CENTURY PLANT

Agave americana L. var. *marginata* Trel.
mecoztli, maguey amarillo, Aztec; *mezcal,* Ópata

Agave atrovirens Karw. ex Salm.
metl, maguey, Aztec; *maguey,* Spanish

Agave bovicornuta Gentry
lechuguilla, Pima Bajo; *sábila,* Tepehuan

Agave parryi Engelm.
kuu chowa, mezcal, Mayo

Agave parviflora Torr.
ma'i, Mountain Pima

Agave spp.
yel, Paipai; *maguey,* Spanish; century plant, English

Agave species are common succulents throughout the warmer parts of the New World. They have a rosette of large, juicy leaves, some of which contain chemicals irritating to the skin. In some species, the leaves can be several feet long and several inches thick. The plants generally live for several years (although not a century), then produce a tall flower stalk, whereupon they flower, set seed, and die. *Agave* gets its botanical name from the Greek word for noble, referring to the tall inflorescence of *A. americana.*

Historic Use. "Almost innumerable are the uses of this plant," declared Hernández (1959, 2:348): it was thought to bring down menstruation, soothe the stomach, provoke urine, clean the kidneys and bladder, break up stones, and clean the ureters. The juice, which was said to be 'cold' and 'moist', becomes glutinous when roasted. The roasted leaves when applied were said to cure convulsions and calm the pain of the "Indian plague," especially if the drink was taken hot. It would dull sensitivity and cause drowsiness. The juice of three or four leaves of *maguey amarillo* to which was added three peppers was thought to evacuate, little by little, cold and raw humors from below and via urine. The Aztecs gave this plant to women for a few days after they gave birth to strengthen them. It also was thought to alleviate asthma.

Agave

1. Used for wound healing and as a drink.

2. Contains polysaccharides; is antibiotic, fungistatic, antiinflammatory, estrogenic, and high in vitamin C.

3. Sap can cause dermatitis.

In northwestern New Spain, Esteyneffer ([1719] 1978) repeated these numerous uses of *maguey,* recommending it in addition for dirty or bad-smelling sores, kidney pain, abscesses, wounds, simple ulcers, and sessile warts, as well as for making plasters and evacuating phlegm. The roasted leaves were even said to cure wounds from bull horns or firearms. Gilij (1785) described its use for treating gangrene. Clavijero (1787:53), quoting del Barco, said *mezcal* was similar to *maguey* and was a member of the aloe family, again confusing these two plants. He said that in Baja California, some took it to become intoxicated and others for medicine since it was considered a diuretic and good for the stomach. Pfefferkorn ([1794–95] 1989:60) said that scurvy could be cured in a few days by drinking a small glassful of *mezcal* each morning on an empty stomach. The leaves were to be roasted in hot ashes, the juice pressed out and cooked while being skimmed, then left out to cool. To stop the vomiting of an upset stomach, spirits were distilled from the huge root, which was roasted all night in a pit lined with stones. A small swallow was to be taken every day one hour before the noon and evening meals. The root healed fresh wounds and also old wounds when applied with a cloth saturated with the juice.

Modern Use. The Pima Bajo use the sap from leaves of *lechuguilla* mixed with grease, applying it to the back for kidney problems. The Tepehuan, who are reported as calling the plant *sábila* (the Spanish common name for aloe), put the sap on the cheek for toothache. The Mountain Pima bake the leaves of *A. parviflora* and apply this to wounds. The Mayo roast the *penca* (leaves) *de mezcal,* express the fluid, and take up to two teaspoons for scorpion bite. They cook the heart until the water is like a thick syrup. This is good for a cold. The Paipai toast leaves of *yel* for several hours, then boil the toasted leaves with water. Women drink this tea every morning during the period of the menstrual flow. Mexican Americans in Tucson use the leaf to treat diabetic infections: the pulp of a leaf of *maguey* is applied to the infected area, alternating with aloe.

Phytochemistry. Many of the folk uses stress wound treatment, and wound healing can be effected through agaves' polysaccharides, which are bactericidal, and saponins and sapogenins, which have antibiotic, fungistatic, and possibly antiviral activity (Ortiz de Montellano 1990). *Agave sisalana* Perrine (mescal) contains tigogenin, hecogenin, gitogenin, neo-tigogenin, sarsapogenin, sisalogenin, gloriofenin, gen-trogenin, delta 9-11 hecogenin, diosgenin, and yamogenin pectin, along with much vitamin C (Duke 1985:22). The last mentioned may account for the use of related species by Ópata in the eighteenth century as an antiscorbutic. Be-cause of the saponin content of *Agave schottii*, it was being investigated for cancer treatment (Lewis and Elvin-Lewis 1977).

■ *Allium* (Liliaceae)

GARLIC, ONION

Allium cepa L.
cebolla, Spanish; onion, English

Allium sativum L.
ajo, Spanish; garlic, English

These Old World herbs have been introduced to all parts of the New World, where they are cultivated for food and medicine.

Historic Use. Garlic was well known to the ancient Egyptians, who elevated it to the rank of a deity. The Israelites introduced it to Palestine, the Greeks and Romans used it, and the Crusaders are thought to have taken it back to France (Schauenberg and Paris 1977:84; Font Quer 1979: 888). Dioscorides recommended garlic as an antivenom, vermifuge, and emmenagogue, and Galen said it was 'sharp', 'dry', and 'hot' in the fourth degree. He described various kinds of onion, good for whatever might ail one. In the eighteenth century, Esteyneffer ([1719] 1978) recommended both garlic and onion for a wide variety of illnesses, uses that continue today.

Modern Use. The Mountain Pima insert a clove of garlic into the anus to treat fever. The Mayo use garlic in capsules for arthritis and cancer and to reduce fat. Mexican Americans insert the clove to treat *susto*, fright disease. Both respect garlic for its magical properties. Mexican Americans

Allium

1. Garlic is used for high blood pressure and against fright; onion has been used for fever.

2. Contains a disulfate and vitamins and has antibiotic activity.

3. The essential oil of onion causes tears; onions can cause anemia, jaundice, and digestive disturbances.

treat high blood pressure by eating the clove, followed by a glass of milk, twice daily in addition to taking lemon four to five times daily. They recommend eating garlic for diabetes. They also apply mashed garlic to an insect bite to relieve pain. They grow garlic as an herbicide with roses. Onion, with fewer uses, was formerly fried and applied to the chest for cough and fever.

Phytochemistry. Garlic—especially the bulb—contains allicin and allistatin (which are active against broad-spectrum bacteria and fungi); essential oils consisting of allicin and diallyl sulphide; enzymes including alliinase; and vitamins A, B1, B2, and nicotinamide. The oil is irritating to the skin and mucosa. Both garlic and onion show experimental hypoglycemic activity. Onion contains various sulphides—especially allyl propyl disulphide—flavones, enzymes, and vitamins A, B1, B2, C, and D (Lewis and Elvin-Lewis 1977; Schauenberg and Paris 1977; Font Quer 1979). The essential oil of onion causes tears. Eating large amounts of onions over time will cause anemia, jaundice, and digestive disturbances (Fuller and McClintock 1986:266).

■ *Aloe* (Liliaceae)

ALOE

Aloe barbadensis Mill.
aloë, Renaissance English; *áloe yacapichtlense, tlailochtía, medicina repelente,* Aztec; *acíbar, zadiva,* Ópata; *sábila, zábila,* Spanish; aloe, aloe vera, English

Aloe species are superficially similar to *Agave* species; hence they are often confused. The leaves of aloes are more juicy and less fibrous.

Historic Use. *Aloe,* native to Africa, from where the generic name comes, was early introduced to Barbados and is so extensively cultivated there that the species is botanically named *Aloe barbadensis.* When Columbus first stepped ashore in the New World and was looking for medicinal spices, he thought he saw *Aloe* but in fact was looking at *Agave.* Hernández (1959, 2:149) was not similarly confused; he separated *Agave* from *Aloe,* which had reached Mexico shortly after the Conquest (Valdés and Flores 1984:86). He said the Aztecs used *Aloe* for various skin problems—erysipelas, sores, and infections—for it was "mucilaginous and salivatous."

In the Old World, Gerard ([1633] 1975:508), quoting Dioscorides, said that the juice was good for many uses in physic (medicine). He described aloe as 'hot' in the first degree but 'dry' in the third, extremely bitter. It "purges the stomach, bringing forth choler, preserves dead carcasses from putrefying, purges worms from the belly, brings down the monthly courses, but is not convenient for the liver. It heals green wounds, cleanses ulcers and cures such sores as are hardly to be helped, especially in the fundament and secret parts." Gerard also noted another plant called *Aloe* that grew in America, but he described *Agave*.

Segesser made a similar error with *Agave* in the American and Mexican West during the eighteenth century. Esteyneffer ([1719] 1978) recommended *acíbar* against worms in the ear and in the intestines and, when applied roasted, for callouses and to reduce swelling of the arm or leg after a bleeding treatment.

Modern Use. The Mayo press the juice from aloe and apply it to a boil to make it burst. Yaqui successfully treat bedsores with aloe. Tarahumara treat sores and burns with the juice of the leaf of aloe. Mexican Americans in Las Cruces, New Mexico, treat diabetic infections by removing the skin of a leaf, cutting it in half lengthwise, and after heating it briefly over the kitchen stove, applying the oozing mass to the sore area. They also strip off the epidermis and then squeeze the liquid directly onto a burn. They take three tablespoons of the juice in orange juice each morning for joint pain. They also apply the mucilage to remove the brown spots of aging from their hands. A Tucson informant recounts that when she has colitis or gastrointestinal disturbances, she puts the juice of a leaf of aloe in a glass of water and drinks it to stop the symptoms. She triumphantly tells of curing dermatitis with a mixture of aloe, avocado pulp, and vitamin E cream.

Phytochemistry. *Aloe* contains anthraquinone glycosides, principally barbaloin (up to 30% in *A. barbadensis*). It should not be used in pregnancy or by individuals with hemorrhoids or kidney problems; it has the toxin arabinose

Aloe

1. Used as a cathartic and to heal wounds and infections.

2. Contains up to 30% anthraquinone glycosides and thus would show laxative action.

3. Ingestion is contra-indicated in pregnancy, hemorrhoids, and kidney problems.

(Duke 1985:32, 547). Fluid from fresh leaf sources promotes the attachment and growth of human cells and enhances the healing of wounds (Tyler, Brady, and Robbers 1981:23). It also contains resinous material and volatile oil (Tyler, Brady, and Robbers 1981:62).

■ *Ambrosia* (Asteraceae)
RAGWEED

Ambrosia acanthicarpa Hook
[= *Franseria acanthicarpa* (Hook) Coville]
u'rí, Tarahumara; *fiate*, Warijio

Ambrosia ambrosioides (Cav.) Payne
[= *Franseria ambrosioides* Cav.]
tuguiro, juigiro, Ópata; *chícura, čikuli*, Tepehuan; *chícura, yerba de la muela*, Mountain Pima; *chikuri*, Tarahumara; *chícura*, Baja California Sur; *tincl*, Seri; *yerba del sapo* (frog's herb), New Mexico Spanish; *chícura*, Spanish; canyon ragweed, bursage, false cocklebur, English

Ambrosia psilostachya DC.
chipúna, Tarahumara

Ambrosia spp.
ambrosia, oke of Cappadocia, Renaissance English; *chícura*, Pima Bajo, Mayo

Bursage is a small, aromatic shrub with coarse, bluish green leaves that grows near water and arroyos. It is one of the most common plants in the arid southwestern United States. *Ambrosia* was named from Greek mythology: the food of the gods, which was supposed to be especially delicious or fragrant. Species found in the American and Mexican West share variations of the Spanish name *chicura*.

Historic Use. In the Old World, the plant name *Ambrosia* was listed by Dioscorides. According to Gerard ([1633] 1975 :1109), *Ambrosia* was 'hot' and 'dry' in the second degree and was to be "boyled in wine, and ministered unto such as have their brests stopt, and are short winded, and cannot easily draw their breath." Gerard called it oke of Cappadocia, illustrating it with a drawing of a plant that resembles the New World *Ambrosia*.

Ragweed was used medicinally throughout the American and Mexican West, frequently to treat women's conditions. The first mention was by Esteyneffer, who wrote in the first edition of his work—printed in Mexico in 1713 (1887:70)—of *"chicura (segun llaman en Sonora),"* recommending it as a fumigant for "spasms and convulsions." By 1719 in Amsterdam the name was misprinted as *cicuta* (hemlock), a serious error perpetuated in the editions that followed. It seems likely that the Spanish name, *chicura,* comes from the Ópata *juigiro,* as reported by Nentuig ([1764] 1977): "Its leaves, heated in the hot ashes, and placed on the belly, cures women's hysteria, among its many virtues." Tamayo (1784), in Arivechi, Sonora, recommended cooking the root and placing the decoction on the *yadtras* (the bottom), while the leaf was believed good for "tumors." The branches were burned to ash to make soap.

Modern Use. The Tepehuan give a tea made from the leaves of *čikuli* to women who are experiencing difficulty in parturition. They heat the leaves of a related species and apply them as a poultice. The Pima warm the leaves of *A. ambrosioides* and spread them on the chest to loosen a cough and also have learned to crush and boil *chicura* roots and administer the decoction "for women's pains and menstrual hemorrhage." The leaves are warmed and put on rheumatic pains: "Then you get good feelings," said a Tohono O'odham informant. The Mountain Pima make a tea of the leaves and stems to give before, during, and after childbirth. The Pima Bajo crush the leaves of *chícura* and place them on the head for headache.

The Warijio make an infusion of *A. acanthicarpa* for stomach troubles and colds. The Mayo cook the leaves of *chícura* in water for a tea used for cleaning out the uterus after giving birth. The Tarahumara use the leaves of *A. acanthicarpa* as a poultice for wounds and swellings. They use the leaves of *A. psilostachya* in a tea for gastrointestinal problems. The Seri say that a tea made from the roots of *A. ambrosioides* may be given to a woman after parturition or before.

Ambrosia

1. Used worldwide for women's conditions.

2. Some species have strong antibiotic activity.

3. Contains alkaloids with carcinogenic potential. Flowers are allergenic.

The Zuni (Stevenson 1915:52) make a tea of *A. ambrosioides* that is drunk warm for "obstructed" menstruation. The tea is also rubbed over the abdomen. It is believed to produce abortion if sufficiently strong.

Mexicans of Baja California Sur use a decoction of the leaves of *A. ambrosioides* for headache, colds, rheumatism, and varicose veins and a decoction of the root to fortify the uterus after childbirth and as an abortive. It was used by New Mexico Spanish-Americans to hasten childbirth (van der Eerden 1948:16). Finally, Mexican American women say that they make a douche from the leaves "to clean everything out" after menstruation and also take a tea made of the leaves for menopausal symptoms. The herb is sold in Tucson, though women could collect their own, since it grows abundantly near water.

Phytochemistry. According to NAPRALERT, the aerial parts of *A. ambrosioides* contain numerous cytotoxic compounds: the sesquiterpenes damsin (up to 00.81%), damsinic acid, franserin, parthenolide, psilostachyin, and the flavonoid hispidulin. Sesquiterpene lactones are highly allergenic; ragweed is infamous for inducing allergic responses. Apparently no studies have been conducted to explain its frequent use for conditions of women, although one might postulate that the cytotoxic compounds contribute to "cleaning everything out." Compounds in the root have not been identified. An extract of the branches of *A. ambrosioides* had 2 + activity against *Staphylococcus aureus* and 1 + against *Bacillus subtilis,* while an extract of the whole plant of *A. psilostachya* had 2 + activity against *Staphylococcus aureus, Bacillus subtilis,* and *Streptococcus fecalis.*

■ *Anemopsis* (Saururaceae)

LIZARD TAIL

Anemopsis californica (Nutt.) Hook & Arn.

> *guaguat,* Ópata; *vavish,* Pima; *babis, hiervelmanso,* Mayo, Yaqui; *cha pan,* Paipai; *hierba de manso, comáanal,* Seri; *yerba del manso, hierba el manso, yerba mansa, mansa, bavis, bavisa,* Spanish; lizard tail, manso grass, English

Yerba mansa is an herbaceous plant native to wet areas in the mountains of the southwestern United States and Mexico. It has a large, dense spike of small flowers with white bracts. There are more entries in my data assemblage for *A. californica* than for any other species in the American and Mexican West. It is used as well by groups outside the delineated area, by the Chumash (Timbrook 1987), Costanoan, Kawaiisu, Mahuna, Paiute, and Shoshone (Moerman 1986:36).

Historic Use. *Anemopsis* received its botanical name from its supposed resemblance (*-opsis*) to the anemone. Pliny (1938, 6:279) recommended anemone—which is an entirely different plant—for headache and inflammations, uterine complaints, and troubles with lactation, as well as for promoting menstruation, cleansing ulcerous sores, and healing teeth. Gerard ([1633] 1975:374) noted numerous kinds of anemones and made the same recommendations as Pliny.

In northwestern New Spain, we first hear of *hierba del manzo* from Esteyneffer ([1719] 1978:632, 633, 635), who reported that it was used to stanch bleeding from wounds and bruises as well as for problems with teeth. Nentuig ([1764] 1977:61) said that the Ópata held a decoction of *hierba del manzo* in the mouth for a toothache and used the plant for many of the same problems for which the Old World anemone was employed. It was taken as a tea to relieve anxiety and anguish from nervous attacks. Fried in tallow or reduced to a powder, it was applied directly to tears caused in childbirth.

Modern Use. The Pima make an infusion from the roots for application to sores and also take it internally to treat syphilis. Both the Pima and the Tohono O'odham take a hot tea from the roots of *Anemopsis* for cold and cough. Afterward the patient is covered with blankets to cause sweating. A piece of dry root is held in the mouth to prevent cough. The Yaqui boil the root to make an infusion to wash infected sores. The Mayo cook the leaves in water with salt: with this they wash an infected cut three times daily, then dry the wound and place a few leaves in it. They also drink the tea to treat cancer of the uterus and use the water as a douche to treat excessive bleeding after childbirth.

Anemopsis

1. Widely used for infected sores, colds, and women's health problems.

2. Believed to be analgesic, antiseptic, fungicidal, and sedative.

3. Toxicity not reported.

The Paipai of Santa Catarina use a brew for childbirth wounds: they grind the root and boil the resultant powder into a tea for colds (Owen 1963:338). The Seri are reported to make a tea with it and *yerba buena (Mentha)* to induce conception. The legs are bathed with another mixture including *comáanal* to treat rheumatism. A decoction of the whole plant including the roots is used to disinfect and cure sores. The sores are washed with the brew, or a cloth soaked in the liquid is placed on them. The Seri also hold tea from the root in the mouth over an aching tooth. Seri knowledge of this plant was reportedly obtained from a Mexican woman.

Mexicans make a powder from the dry roots for a tea or a wash for healing wounds or skin disorders. The Mexicans of Baja California Sur make a tea from the leaves for a cold. They also use this tea for stomachache, leprosy, and problems with the kidneys and blood circulation. They bathe wounds or bruises with the tea or apply it on a poultice.

Mexican Americans in Las Cruces, New Mexico, who call the plant *bavisa,* make a brew from the leaves to rinse infected areas. To prevent infection of a burn, they leave the area exposed to dry and then apply cold milk or ice to the area. They drink the tea for menstrual pain and, with lemon, for relief of nausea. They also use it for toothache. A cup of *bavisa* tea with sugar is taken for stomachache, an internal infection, or *pasmo.* For cough, they make a brew, adding honey, lemon, and Vick's cough drops. A rinse made from the root treats swelling. For pain, they soak the affected joint in warm tea and also drink the tea. A *bavisa* decoction is used as a vaginal douche. The brew is also taken for cancer of the uterus. It is well known as helpful in "women's problems." *Yerba mansa* is sold by an herbalist in Mesilla, New Mexico, in Douglas, Arizona, and in a Tucson *botanica,* but it can be harvested near water.

Phytochemistry. According to NAPRALERT, *A. californica* contains eugenol in the rhizome and root. Estragole, thymol, methylether, linalool, p-cymene, asarinin, and other aromatics have also been found (Moore 1989:133). Interestingly, in one study no antibiotic activity was found in an extract of the whole plant (Encarnación and Keer 1991).

■ *Arctostaphylos* (Ericaceae)

BEARBERRY, UVA URSI

Arctostaphylos uva-ursi (L.) Spreng.

Arctostaphylos pungens H.B.K.
yori, Mountain Pima; *wichari*, Tarahumara; *manzanilla*,
uhi, Warijio; *pinguica*, Baja California Norte; *pingüica*,
Spanish

Arctostaphylos spp.
barberry, *spina acida, sive oxyacantha*, Renaissance English; *tepetómatl* (mountain tomatoes), *tomázquitl*,
planta que da frutos acinosos parecidos al tómatl, Aztec;
manzanita, Tepehuan, Spanish, English

Manzanitas are evergreen shrubs with crooked, twisted
trunks and branches and reddish bark, growing at altitudes
of 3,000 to 9,000 feet. The fruits are small spherical berries
eaten by many kinds of wildlife, including bears, from
which come the botanical and English names. *Manzanita*
means "little apple."

Historic Use. The great Greek and Latin pharmacologists
apparently did not know the bush. The first report of it
appeared in the thirteenth century in an English book of
medicinal plants. Clusius in 1576 described the plant in
Spain (Font Quer 1979:536–37). Gerard ([1633] 1975:1325–
26) proclaimed that the leaves and berries were 'cold' and
'dry' in the second degree and that Galen noted that they
were of cutting quality. The decoction was said to be good
against "hot burnings and cholericky agues."

Hernández (1959, 2:232) classified manzanitas as species
of madrone. He said that washing with the decoction reduced swelling of arms and legs, while drinking it was astringent to the belly. Drinking a decoction of the roots of
another species (which "gave acid fruits resembling tomatoes") that was smaller, and had smaller fruit, for ten con-
secutive days was beneficial to asthmatics.

Modern Use. The Mountain Pima make a drink from manzanita leaves to treat kidney pain. The Tepehuan make a tea
from the leaves to relieve the discomfort of a cold. The
Tarahumara make a wash for sores and a tea for chest and

Arctostaphylos

1. Used for bladder and kidney conditions.

2. Contains an effective urinary antiseptic, hydroquinone.

3. Hydroquinone in large quantities can be toxic. Tannins can upset the stomach.

lung conditions. The Warijio eat the berries and report that an infusion is made from the leaves for colds and *sarampión* (measles). In Baja California Norte, *A. pungens* is recommended to treat renal problems, gall bladder problems, urinary tract infections, nephritis, and prostate problems. None of my informants described using the dried fruit, but it is sold in *yerberias* and supermarkets catering to Mexican Americans. It is an ingredient in most kidney and bladder teas available in health food stores.

Phytochemistry. Biochemical investigations are lacking for *A. pungens* (Winkelman 1986). However, *A. uva-ursi* contains 5–18% arbutin, 6–40% tannin, allantoin, various acids, and isoquercetin (Duke 1985:55–56). Arbutin hydrolyzes to yield hydroquinone, an effective urinary disinfectant if the urine is alkaline. In very large doses hydroquinone is toxic, leading to collapse and possibly even death. Large doses are oxytocic. To minimize the tannin content, which can upset the stomach, leaves should not be extracted with hot water (Tyler 1993:313–14).

■ *Argemone* (Papaveraceae)
PRICKLY POPPY

Argemone mexicana L.
chicálotl, Aztec; *chicalote, cardo santo campestre,* northwestern New Spain; *hipicdum,* Pima Bajo; *tachina, chicacotl,* Mayo; prickly poppy, holy thistle, English

Argemone ochroleuca Sweet
hierba loca, Mountain Pima; *tachiná,* Warijio; *cardo,* Tepehuan, Spanish

Argemone platyceras Link & Otto
XaSácöz, Seri

Argemone spp.
argemone capitulo torulo, bastard wilde poppy, Renaissance English

Prickly poppies are tall field and roadside weeds with attractive white, yellow, or orange flowers but stiff, sharp spines. The leaves and stem contain an acrid yellow latex.

Historic Use. This plant is an American native. However, in the Old World, according to Gerard ([1633] 1975:373), an *Argemone* was known and classified as 'hot' and 'dry' in the third degree. He said the plant took its name from a disease of the eye called argema, which it cured. Pliny (1938, 7:283) recommended argemonia for quinsy (sore throat), spleen, gouty pain, inflammations, carcinoma, and warts.

The Aztecs used this poppy primarily for inflamed eyes, wrote Hernández (1959, 2:267), who noted its resemblance to *cardo santo* or holy thistle, but it had many other uses as well. They mixed it with the milk of a woman who had recently delivered a girl. The ground seeds in a two-dram dose were said to be good to evacuate all humors (especially those produced by the pituitary, which were believed to damage the joints) and as a purgative and also for fever and ulcers on the genitals. Applying the flower was thought to cure skin disease. Mixed with mesquite sprouts it was believed to dissolve cataracts. Hernández proclaimed it bitter, 'hot', and 'dry'. The only mention in northwestern New Spain was Esteyneffer's recommendation of using half a tablespoon of toasted and ground seed in a cup of water or broth as a purgative ([1719] 1978:744).

Modern Use. The Mayo grind the seeds of *A. mexicana*, mix in warm water, and drink the mixture for a purgative, as do the Pima Bajo; and the seeds of *A. ochroleuca* prepared the same way in a drink are used by the Tepehuan. The Warijio apply the juice of *A. ochroleuca* to treat sore eyes. The Mountain Pima boil the leaves to wash wounds. If the placenta is not seen to be complete, the Seri mother is given a tea of *A. platyceras* to cause the expulsion of the remaining portion. The Seri make a tea with *A. mexicana* to treat urinary problems. The Tepehuan apply the milky excrescence from the stalk of *A. ochroleuca* to kill fleas or lice. This may be the only safe use for *Argemone*. However, the juice has been recommended for topical application to burns and scrapes and a drink of the tea for sedation and to relieve palpitations and nerve pain (Moore 1989:91–93). Mexican Americans in Las Cruces, New Mexico, cut the flower, gather the brown liquid on cotton, and add it to boiled water, which they cool and use as eye drops morning and night for *punzadas en el ojo*, pain in the eye.

Argemone

1. Used as eye drops and as a purgative.

2. May have antimicrobial and analgesic properties.

3. Is narcotic, hallucinogenic, and toxic: seeds have caused fatalities, oil causes edema and glaucoma.

FIGURE 10 *Argemone
platyceras.* a. Flower
bud. b. Seedpod. c. Seed.

Phytochemistry. *Argemone mexicana* has been reported to
contain several alkaloids including alocryptopine, berberine, chelerythrine, coptisine, dihydrochelerythrine, protopine, and sanguinarine (Duke 1985:59). The flowers contain glucosides, and the seeds contain 20–35% oil (which
causes edema and glaucoma). The plant is narcotic, possibly hallucinogenic, and the seeds have caused many fatalities. Toxins include biflorine, protopine, ricinoleic acid,
sanguinarine, and possibly codeine and morphine. Smoking the seeds can cause nausea, vomiting, and diarrhea.

■ *Aristolochia* (Aristolochiaceae)

BIRTHWORT, SNAKEROOT

Aristolochia brevipes Benth.
santa maría, Tarahumara; *cosa naajiburia, gozanajibi-
juria,* Mayo; *hierba del indio,* Pima Bajo, Yaqui, Baja
California; birthwort, snakeroot, English

Aristolochia mexicana Kostel.
phehuame, medicina buena para el parto (good birth
medicine), Aztec

Aristolochia pentandra Jacq.
yerba del indio, Baja California

Aristolochia quercetorum Standl.
hierba del indio, Warijio

Aristolochia watsonii Woot. & Standl., and *Aristolochia
brevipes* Benth.
hataast an ihiih (what gets in teeth), Seri

Aristolochia wrightii Seem.
yerba del indio, Tarahumara

Aristolochia spp.
guaco, northwestern New Spain; *yerba del indio,* Mayo,
Spanish

Aristolochia is a genus of woody climbers, most of which
are tropical but a few of which occur in temperate regions.
The flowers are purple, shaped something like a large
smoking pipe (Mason and Mason 1987:66), and foul-
smelling. The botanical name *Aristolochia* comes from the
Greek *aristos,* best, and *lochia,* childbirth. Most of the ethnic groups in the American and Mexican West name all the
various *Aristolochia* species *yerba del indio,* Indian herb.

Historic Use. Gerard ([1633] 1975:848–49) reported various
species that were noted by Pliny, Dioscorides, and Galen,
adding *Aristolochia virginia* as a new species with all the
common uses of the others but also able to serve as rat-
tlesnake antidote. Hernández (1959, 3:91) said that the Aztec
Aristolochia (mexicana) was the same as "*Aristolochia cle-
matis* by our Dioscorides" and recommended a decoction

Aristolochia

1. Used as a douche or taken for stomachache.

2. Is antiseptic, bacteriocidic.

3. A local irritant—extended use or too large a dose can cause coma and death.

of the root bark for asthma and any affliction coming from 'cold', such as chronic cough, kidney stones, the "Spanish plague" (syphilis), and headache. It was thought to accelerate birth, provoke menses, and open obstructions in women by getting rid of the 'cold'. The herb's qualities were odorous, sharp, 'hot', and 'dry' in the third degree. Hernández said that since it grew so well in Michoacán, it could easily be transplanted to Spain.

In northwestern New Spain, Esteyneffer ([1719] 1978) said that *guaco* root, when dried and pulverized, healed and prevented scarring of venereal ulcers and restored the gums. Noting Linnaeus's aphorism that all plants of this genus, whether round or long, have the same virtues, Longinos in Baja California (1792) said that the root, when pulverized and applied, was good for wounds and ulcers (Engstrand 1981:137).

Modern Use. The Pima Bajo boil the root of *A. brevipes* into a potion that they take for kidney or stomach disorders. The Yaqui store the roots of *A. brevipes* for long periods of time for later cooking into a tea for stomachaches. The Warijio use the root decoction of *A. brevipes* for kidney and stomach ailments and *A. quercetorum* for *empacho* and also as a wash for sores. The Mayo toast and grind the root of a *hierba del indio;* mix it with salt, mint, and basil in boiled water; and take one-half cup once a day for *empacho*—or they may simply cook the root in water. The Tarahumara make a tea of the resin of *A. wrightii* for stomach complaints and of the root for fever. Baja Californians Sur take the decoction of *A. brevipes* as a tea for diabetes, fever, malaria, and diarrhea. They say that *Aristolochia* can be applied in a poultice for stomachache, taking seven days to work. The Seri cook the herbage of *Aristolochia* species to make a decoction for toothache or heat the root until dry, then place it over a cavity in a tooth. Mexican Americans in Tucson use the herbage or root of a *yerba del indio* in a decoction as a vaginal douche.

Phytochemistry. Little phytochemical study has been conducted on species in the *Aristolochia* genus despite frequent use in various parts of the world. A series of alkaloidal acids (e.g., aristolochic acid I) has been reported from a variety of species. The root of *A. brevipes* was found to be highly active against *Bacillus subtilis* (Encarnación and Keer 1991). No other studies of the *Aristolochia* species of the American and Mexican West were located. Norman Farnsworth (1993a:36D) has stated that "no herbal medicine in the genus *Aristolochia* should ever be used by humans over an extended period of time." *Aristolochia serpentaria* L. was found to contain gums, resins, aristolochine, an essential oil, and aristolochic acid I among its compounds (Duke 1985:63). Too large a dose of aristolochine is toxic, violently irritating the gastrointestinal tract and kidneys, even causing coma and death from respiratory paralysis.

■ *Arracacia* (Apiaceae)

ARRACACHA

Arracacia atropurpurea Benth. & Hook
acocotli tepecuacuilcense, hierba de tallo hueco (hollow stemmed herb), Aztec; *comino,* northwestern New Spain

Arracacia brandegeei Coulter & Rose
chuchupate, Baja California Sur

Arracacia edulis S. Wats.
yerba del oso, Mountain Pima

These plants have fernlike leaves and small umbels (flowers in a flat-topped cluster resembling a parasol), in addition to carrotlike roots.

Historic Use. The Aztecs used the juice of *acocotli* as a mouthwash for ulcers. They applied the leaves as a plaster or ground the leaves to a powder, which they applied to swollen legs to cure chronic ulcers. The herb was described as glutinous, 'cold', and 'moist' (Hernández 1959, 2:8). Esteyneffer ([1719] 1978) recommended *comino* parts in a decoction, plaster, drink, and bath for twenty-seven different conditions.

Modern Use. In Baja California Sur the root is prepared as a tea that is taken to treat many different conditions: rheumatism, stomachache, kidney pain, high blood pressure, and diabetes. The Mountain Pima wash and soak the seeds

---

Arracacia

1. Has been used for diabetes, high blood pressure, and pain.

2. Other species contain vitamin A and essential amino acids.

3. No reported toxicity.

in water, then drink the fluid for fever. This medicinal plant has not been reported to me by Mexican Americans in Arizona or New Mexico, but it was found to be commonly used in east Texas, where it is called *comino.*

Phytochemistry. An extract of *A. brandegeei* root was found to have 2+ effectiveness against *Staphylococcus aureus* and 1+ against *Bacillus subtilis* (Encarnación and Keer 1991). No other information on the *Arracacia* species listed above could be found. Other *Arracacia* species are reported to contain carotenoids, provitamin A (Bicudo de Almeida and Penteado 1987), pyranocoumarins (Delgado and Garduno 1987), and the amino acids methionine, cystine, lysine, and tryptophan (Norberga 1975).

■ *Artemisia* (Asteraceae)

WORMWOOD, MUGWORT, WESTERN
MUGWORT, SAGEBRUSH

Artemisia annua L.
sweet wormwood, English

Artemisia dracunculoides Pursh.
xal paq, yerba niso, Paipai; *estafiate,* Tepehuan, Spanish; absinth, English

Artemisia franserioides E. Greene
altamisa de la sierra, New Mexico Spanish

Artemisia frigida Willd.
altamisa de la sierra, plumajillo de la sierra, Colorado Spanish; fringed sagebrush, English

Artemisia ludoviciana Nutt.
estafiete, musha, mostafiete del campo, ostafiete del bosque, kokmok sha'i, Mountain Pima; *rosáwari,* Tarahumara; *altamisa de la casa,* Colorado Spanish; *estafiate, romerillo,* New Mexico Spanish; western mugwort, English

Artemisia mexicana Willd.
iztauhyatl, sal amarga, anónima mechoacanense, tzaguángueni, iztáuhyatl latifolio, Aztec; *ajenjo del pais, ajenjo,* northwestern New Spain; *túparo,* Ópata; *istafiate, estafiate,* Yaqui, Mayo

Artemisia tridentata Nutt.

Ts'ah (the sagebrush), Navajo; *chamiso hediondo, chami-
sona,* Spanish; big sagebrush, English

Artemisia vulgaris L.

yerba de san juan, corona de san juan, northwestern New
Spain; *istafiate, romerillo,* Spanish; mugwort, English

Artemisia species range from the small *A. frigida,* less than
one foot high, to the large sagebrush *A. tridentata.* Bluish
green sagebrush is an extremely common shrub characteristic of the higher desert regions of the Southwest. Many
Artemisia species have silvery, faintly hairy leaves on the
underside or both sides, with tiny yellow flowers on spikes.

Historic Use. The name *Artemisia* has through history
mixed a variety of genera and species, primarily Asteraceae,
including absinth, *Ambrosia,* chrysanthemum, feverfew,
Parthenium, and *Senecio.* Pliny (1938, 7:381) in the first century A.D. and Claude Lévi-Strauss (1966:46) in the twentieth century noted that *Artemisia* was used in the popular
pharmacopoeia everywhere for women's conditions, to
regulate menstruation and to stimulate childbirth. Thus,
the generic name may derive from the Greek goddess of
women, Artemis. Cones of *A. vulgaris* have been used for
many centuries in Japan, China, and elsewhere to burn on
acupuncture points, for moxibustion.

Hernández described three plants used by the Aztecs
that he declared to be equivalent to European *ajenjo,*
Artemisia. They were categorized as bitter, 'hot', and 'dry' in
the third degree. The Aztecs used *Artemisia* for colic, "to
take away pain that comes from cold or flatulence." It was
said to relieve infants who vomited or had *empacho.* The
decoction was used as a wash on the legs to help swelling
(Hernández 1959, 3:7). Another *Artemisia, tzaguángueni,*
when applied to hemorrhoids dried and constricted them
(Hernández 1959, 2:189). The Aztec name *iztauhyatl* was
hispanicized to *istafiate* or *estafiate,* today's Spanish name.

In northwestern New Spain, Esteyneffer ([1719] 1978)
recommended *Artemisia* for 28 conditions, including hemorrhoids, urinary retention, gonorrhea, childbirth illness,
and wounds. Nentuig ([1764] 1977:61) noted that *ajenjos o
estafiate, túparo* was one of the herbs already known to

Artemisia

1. Used for women's conditions and gastrointestinal
disorders.

2. Constituents may in-
clude estrogens, affecting
female physiology.

3. Camphor tannins and
thujone can cause epileptic
spasms. Pollens may cause
allergies.

medicine. The powdered root of an *Artemisia* was given in food in cases of rabid animal bites. It was fried and applied to the stomach for disorders (Pablos 1784). In colonial Mexico—for religio-political reasons—*Rosmarinus* was adopted as a generic substitute for the Aztec *Artemisia* (Ortiz de Montellano 1990:205); however, this seems not to have been the case in the American and Mexican West. Although both rosemary and *Artemisia* are used for women's conditions, *Artemisia* occupies a far larger place in the folk pharmacopoeia.

Modern Use. At least three species of *Artemisia* are used interchangeably in the American and Mexican West. Yaqui fry *istafiate* in lard or grease, then apply the ointment as a plaster to the stomach for *empacho* or other stomach disorders. The Mayo cook the root in water, drinking the tea for nine days to treat diarrhea or cachexia. The Tepehuan take it as tea for colic pains and for cold. The Pima Bajo make a tea from the leaves for stomach disorders. The leaves of *A. ludoviciana* are used in a tea for menstrual pains and for gastrointestinal disorders by the Tarahumara. The Mountain Pima, who have various names for *A. ludoviciana*, make a tea with the leaves and sometimes the root for stomachache (Laferrière, Weber, and Kohlhepp 1991). The Paipai also use *A. dracunculoides* for stomachache.

Mexican Americans in Colorado (whose name for *A. ludoviciana—altamisa—*resembles *artemisia*) chew and swallow the dry leaves for stomachaches, especially those "caused by cold." It is an ingredient in tea for diabetes. A bath is made from an infusion of the foliage for fever. The infusion is drunk for colic pain. *Artemisia tridentata*, sagebrush, is used in tea or hot vapor bath for colds. New Mexico Spanish-Americans use *A. franserioides* for abortion, amenorrhea, menstrual pain, and the menopausal hot flash. They use *A. tridentata* in an external poultice over the umbilical stump of the baby to prevent infection and on the small of the back for menstrual cramps and to heal and prevent infection after miscarriage. Mexican Americans in southern Arizona and New Mexico use *istafiate* after childbirth. The leaves of the plant are brewed with baking soda and salt in a tea taken twice daily for six weeks to decrease swelling. The tea is given for menstrual cramps and for symptoms of menopause. A tea made from only two leaves (because the brew is bitter) is given to a child with diarrhea or one with *empacho,* an intestinal infection.

The Navajo use *A. tridentata* extensively, curing headaches by odor alone or boiling the plant to help in childbirth or for indigestion and constipation; they use a tea made from stems and leaves for colds and fevers and a poultice made from pounded leaves for colds, swellings, tuberculosis, and corns (Mayes and Lacey 1989:106–7). Their neighbors the Hopi, Tewa, and Zuni also make medicinal use of sagebrush and other *Artemisia* species. The Dena'ina, Athabascans in Alaska, use *Artemisia* in similar ways: in steam baths for pregnant women, as poultices on the stomach, as a wash for various infections, sore eyes, and arthritis (Kari 1991:139). *Artemisia tilesii* has been called the panacea of the Eskimo peoples, Inupiat, Yupik, Koniag, and Aleuts (Fortuine 1988:215).

Absinth was a favorite recreational drug in the last century, both in Europe and the United States, and its use is being revived today.

Phytochemistry. *Artemisia annua* is the source of the antimalarial drug artemisian. *Artemisia vulgaris* contains camphor, tannins, and thujone, which in large doses can cause epileptic spasms. Cineole, an expectorant and insecticide, is the major constituent of *A. vulgaris* (Duke 1985:70). It also contains stigmasterol, beta-sitosterol, and alpha- and beta-pinene, which are expectorants (Duke 1985:70). *Artemisia dracunculus* produces an essential oil that contains phellandrine, ocimene, and methylchavicol, as well as the hydroxycoumarin herniarin (Schauenberg and Paris 1977:222). The species also contains rutin, which decreases capillary permeability and fragility and is said to be a cancer preventive (although estragole, the main constituent of the essential oil, is reported to produce tumors in mice; Duke 1985:68). The astringent effects of *A. tridentata* likely come from the tannins, and other therapeutic effects, including antibacterial and antifungal effects, from its volatile oil constituents (Brinker 1991–92). As with any Asteraceae, the pollen may cause allergies.

■ *Asclepias* (Asclepiadaceae)

MILKWEED

Asclepias albicans S. Wats.
najcáazzjc, Seri; *mate candelilla, yamate,* Spanish; white
   stem milkweed, English

Asclepias asperula (Decne.) Woods
inmortal, New Mexico Spanish; spider milkweed, ante-
   lope horns, English

Asclepias atropurpurea
[= *Asclepias atroviolacea* Woods?]
oreja del venado, Tepehuan

Asclepias coulteri A. Gray
wichore, Mayo

Asclepias hypoleuca (A. Gray) Woods
suimali nanaaka (venado orejas), Tepehuan

Asclepias lemmonii A. Gray
inmortal, Mountain Pima, Spanish

Asclepias linaria Cav.
tlalacxoyatl, Aztec; *áli okága,* Tepehuan

Asclepias sp. *quinquedentada* A. Gray
contra yerba de la sierra, Tepehuan

Asclepias subulata Decne.
najcáazjc, Seri; *jumete, yamete,* Baja California Sur, Spanish; reed-stem milkweed, English

Asclepias spp.
asclepias flore albo, white swallow wort, Renaissance English; *chupi,* Ópata; *yerba del hígado de flores amarillas,*
   Tepehuan; *yerba del hígado,* Mountain Pima; *tcomalí-
   naka, chiwainame,* Tarahumara; *inmortal, hierba del
   indio,* Tepehuan, Spanish; milkweed, English

Milkweeds are common plants, most of which have opposite leaves and white milky sap. More than one hundred
species of *Asclepias* have been identified in North America.
The Spanish name *inmortal* reflects the plant's ability to re-
generate annually from the root.

Historic Use. Gerard ([1633] 1975:898) reported that "Asculapius (who is said to be the first inventer of Physick, whom therefore the Greeks and Gentiles honored as a God) called it after his owne name." He quotes Dioscorides as having written "that the roots of Asclepias or Swallow-wort boiled in wine and the decoction drunk are a remedy against the gripings of the belly, the stingings of serpents, and against deadly poison. The leaves boiled and applied in form of a poultice cure the evil sores of the paps or dugs [breasts], and matrix, that are hard to be cured."

The Aztecs used roots of *Asclepias linaria, tlalacxoyatl,* for ulcers (Hernández 1959, 2:14). The Ópata, who called it *chupi,* powdered the root and sniffed it like tobacco through the nostrils to get rid of a headache. It grew in the coldest parts of the province (Nentuig [1764] 1977:63).

Modern Use. The Mayo in Las Animas, Sonora, cook *wichore* in plain water, which they take twice daily for fever. The Seri cook the root of *A. subulata* together with the root of *Encelia* and the entire plant of a small spurge as a drink to remedy heart pain. It is held in the mouth to relieve a toothache. The Pima use *A. subulata* (its common name is not remembered) as a physic and emetic. In Baja California Sur, the latex of this species is applied directly to corns and skin ulcers.

The Tepehuan use various *Asclepias* species. They crush the leaves and stem of *A. atropurpurea* for a poultice of *oreja del venado* to treat backache. They make *A. hypoleuca* into a weak tea to relieve discomfort from overeating. The leaves, soaked in cold water for twenty-four hours, act as a purgative. *Asclepias linaria* leaves are crushed, placed on the temples, and held there with a cloth band for headache. A few leaves are prepared as a tea to induce vomiting for those afflicted with severe stomach cramps. Another species of *Asclepias, contra yerba de la sierra,* was used for fever, while still another, *inmortal,* is made into a tea to alleviate throat discomforts of a common cold.

The Tarahumara take a species of *Asclepias,* pound and scald the roots, strain and drink the fluid, then take a drink of fresh water. The purgative action of this drink is supposed to treat venereal disease. They make a poultice from another species for rheumatism. The Mountain Pima use

Asclepias

1. Root used for colds, kidney problems, sluggish heart, and uterine complaints; latex for teeth, corns, skin ulcers, pain, and fever.

2. Purgative, antibiotic, and cardiotonic.

3. Some species have cardioactive glycosides (at toxic levels) and toxic resins that can cause severe poisoning and contact dermatitis.

several *Asclepias* species. The leaves of one *inmortal* are combined with pork fat and applied to congested nostrils and also to wounds. The leaves of another are prepared as a drink for a cold. The roots of the same plant may be mashed and mixed with cooking oil and boiled for a cough medicine, while roots of *A. lemmonii* are used in a drink to alleviate stomach cramps. The leaves of yet another *Asclepias* species are used in a beverage for liver disorders, and a plant identified as *Asclepias strictiflora* is used in a drink for fever, while Tepehuan children eat the fruit.

Mexican Americans in Tucson use the root of a milkweed in a tea for colds and also for kidney problems. New Mexico Spanish use *A. asperula* latex for fevers, aches, pains, and respiratory complaints. In a mild tea it is used to alleviate "sluggish heart" discomfort (Kelley, Appelt, and Appelt 1988:2). A decoction of the root is taken three to four times in the morning as an emmenagogue or abortifacient. A little of the finely ground root is given in cold water or rubbed on the abdomen for labor pains. A hot decoction of the powder is given to facilitate expulsion of the placenta. *Asclepias* species are employed as medicine by the Navajo and Hopi. In this plethora of uses, purgation is the curing action noted by the various ethnic groups.

Phytochemistry. Compounds that have been identified in *Asclepias* species include cardenolides (cardiotonic steroids), which vary by species: *A. subulata* and *A. linaria* are toxic, having a high cardenolide content. Other constituents include flavonoids—especially rutin and quercetin—sterols, triterpenes, and fatty acids (Kelley, Appelt, and Appelt 1988:3). Researchers have found 2+ activity in the root against *Streptococcus fecalis* (Encarnaciónand Keer 1991).

■ *Baccharis* (Asteraceae)

DESERT BROOM, SEEP WILLOW

Baccharis conferta H.B.K.
quauhizquitli, tepopotli, planta de escobas, Aztec

Baccharis glutinosa Pers.
[= *Baccharis salicifolia* (Ruíz & Pavon) Pers.]
baashoma, Mountain Pima; *chagushi,* Tarahumara; *bachamo, bacomo,* Yaqui; *xa'tam mual,* Paipai; *huata-*

mote, Baja California Sur; *caaoj,* Seri; *yerba del pasmo, batamote, guatamote,* Spanish; seep willow, English

Baccharis sarothroides A. Gray

shooshk vakch, Pima; *bawe jeko, escoba amarga,* Mayo; *casol caacol* (large casol), Seri; desert broom, broom baccharis, English

Baccharis thesioides H.B.K.

paashama hioshgama, Mountain Pima; *batamote del monte, yerba del pasmo,* Warijio

Baccharis spp.

Baccharis Monspeliensium, plowman's spikenard, Renaissance English; *romerillo, sissico, escoba amargosa,* Ópata; *pasmy, ropagónowa,* Tarahumara; *machomo,* Mayo; *batamote,* Yaqui; *romerillón, escoba amarga,* Baja California Sur

Batamote is a shrub up to six feet tall that grows along creek banks in the mountainous areas of the Southwest. *Batamote del monte* is a similar species occurring on drier slopes.

Historic Use. In *Baccharis* we have another medicinal plant seemingly used everywhere, with reports of plants having the same name being used in the Old World as well as the New World. The name comes from the Greek god Bacchus. Gerard quotes mention of *Baccharis* by Virgil, Pliny, and Aristophanes and states that *Baccharis* was so called by the Greeks because of its sweet-smelling root. Gerard ([1633] 1975:790–91) said that sitting over a hot bath made of *Baccharis* roots "mightily voideth the birth and furthereth those that have extreme labour in their childing, causing them to have easie deliverance." It also provokes urine, brings down the "desired sickness" (menstruation), and is applied against the bitings of scorpions or other venomous beasts. The leaves, which are astringent or binding, "stop the course of fluxes and rheums."

According to Hernández (1959, 2:39), the Aztecs made a decoction of the leaves together with other plants as a bath to strengthen women who had recently delivered. To grow new hair for those who have lost it, the leaves were mashed and applied to the head. The mashed leaves were also

Baccharis

1. Used to reduce swelling and other problems caused by *pasmo;* also used externally for wounds and pain.

2. Has antiinflammatory, antispasmodic, and antibiotic actions.

3. Contains diterpenes, which may be toxic.

applied to venomous stings. A decoction of the leaves was taken for earache or toothache, was instilled in the nostrils to provoke sneezing and thus evacuate mucus, and was also seen to be good for urine suppression. Hernández described the root as sharp, bitter, burning, 'hot', and 'dry' in the third degree.

In eighteenth-century colonial Sonora, branches of *batamote* were cooked in water with salt, then applied as a compress for rheumatism (Amarillas 1783). The brew was drunk for rabies (Pablos 1784). Father Miguel del Barco reported in 1768 that the Californios bathed their ailing limbs in this brew. Nentuig ([1764] 1977:64) said that the Ópata washed and ground a good handful, put in an equal amount of boiling water, then beat and stirred the mixture constantly until thick to make a plaster on fractures.

Modern Use. The Warijio take an infusion for digestive afflictions. A strong decoction was applied as a hot poultice to reduce the swelling of wounds and sores. The Mayo toast and grind the plant of *B. sarothroides* to a fine powder and apply it to sores caused by *pasmo* twice daily. A decoction of the plant is given to drink and to bathe in for rabies. The Yaqui place a plug of leaves into the ear to relieve earache. The Tarahumara drink a decoction of the leaves and stem of a *Baccharis* species for cold and cough. They poultice wounds with the decoction to reduce swelling. The Mountain Pima make a solution of *B. thesioides* to bathe bruises and *B. glutinosa* for fever, headache, and earache. The Seri make a tea from the leaves of *B. glutinosa* for contraception and also to help one lose weight. They heat the leaves over coals and place the heated leaves on the head as a remedy for headache. A tea made from the leaves is taken to stop blood loss. They also cook *B. sarothroides* twigs into a tea to make a remedy for a cold, or they rub the tea on sore muscles for relief.

Mexicans of Baja California Sur make a decoction of *Baccharis* species as a bath to treat rheumatism and paralysis. The plants are applied to the painful area; or the branches may be fried in a mixture of gasoline, camphor, liniment, and pig fat, with the resulting ointment applied before the patient goes to sleep. They use a decoction of the root of *B. glutinosa* for loss of hair and for kidney pain. The

FIGURE 11 *Baccharis glutinosa.* a. Branch showing male heads, from male plant. b. Leafy branch from female plant with female flower head in fruit. c. Seedlike achene with tuft of fine silky hairs.

branches are boiled for a decoction to treat wounds and toothache. The Paipai report that the leaves are heated over a fire and applied to the painful area of wounds, bruises, and so forth.

Both *Baccharis* and *Haplopappus* (q.v.) species have been called *yerba del pasmo* because they have been used to treat *pasmo*, a condition characterized by swelling, believed to be caused by rapid temperature change from hot to cold. Mexican Americans use this *yerba del pasmo* in a tea. Children with *pasmo*, which can follow dental work or surgery, are given the herb to inhale in steam because the taste is so bitter. Laferrière (personal communication), who was given this herb by mistake when he was conducting field-work with the Mountain Pima in Nabogame, found it had a bad taste.

Phytochemistry. According to NAPRALERT, among the compounds in *B. glutinosa* are kaempferol and Baccharis diterpenes, luteolins, and flavonoids. Also reported to be present are limonene, which inhibits prostaglandin synthesis and is antiseptic, and myrcene, which is also antiseptic (Ortiz de Montellano and Browner 1985:77). Although NAPRALERT reported no studies by 1990 of biological activity for *B. glutinosa,* branches have more recently been reported to have 3 + activity against *Staphylococcus aureus* and 2 + against *Bacillus subtilis* and *Streptococcus fecalis* (Encarnación and Keer 1991).

■ *Bocconia* (Papaveraceae)

TREE CELANDINE

Bocconia arborea S. Wats.
palo del diablo, Tepehuan; *llorasangre,* Spanish

Bocconia frutescens L.
cocoxíhuitl, hierba acre, Aztec; *llorasangre,* Spanish

Bocconia species are trees up to twenty-five feet in height with large leaves and yellow sap. They grow naturally in equatorial Mexico and are often planted in warmer parts of the United States because of their tropical appearance.

Bocconia

1. Has been used for styes, sores, wounds, and headache.

2. Possibly has diuretic, antimicrobial, and anti-inflammatory activity.

3. Toxins include protopine and sanguinarine.

Historic Use. The Aztecs used *Bocconia* as a diuretic and also to treat inflammations and constipation. Hernández (1959, 2:205) noted that the Indians called the plant an herb even though it generally grew to the size of a tree. He said it was 'hot' and 'dry' in the fourth degree with some astringency, with a bitter and sour taste. The shoots, stripped of bark, were used to dissolve cataracts and cloudy eyes, and applications of the mashed leaves treated old sores and warts, especially on the prepuce and other genital parts. The sap calmed stomach cramps caused by 'cold'.

Modern Use. The Tepehuan use the bark scrapings of *palo del diablo* in a poultice for styes or in a tea for stomach upsets. They crush and moisten the leaves, then apply the leaves to treat sores, wounds, goiter, and headache. This plant was not familiar to Mexican Americans. It is included here because medicinal use of this Aztec plant has disappeared in the American and Mexican West, retained only by the Tepehuan: biomedicines appear to have superseded *Bocconia.*

Phytochemistry. According to NAPRALERT, only the fruit, leaf, stem, and bark of *B. arborea* contain alkaloids. *Bocconia frutescens* has chelerythrine and sanguinarine, aqueous extracts of which would be expected to have diuretic, antimicrobial, and antiinflammatory activity (Ortiz de Montellano 1990:255).

■ *Buddleia* (Buddleiaceae)
BUTTERFLY BUSH

Buddleia americana L.
topozan, zayolizcan, Aztec

Buddleia cordata H.B.K.
matowi, batówi, Tarahumara

Buddleia crotonoides A. Gray
lengua de buey, Baja California Sur

Buddleia perfoliata H.B.K.
salvia real, Baja California Norte; *salvia de bolita,* Spanish

Buddleia sessiliflora H.B.K.
tepozán, Tepehuan, Mountain Pima

Buddleia tomentella Standl.
palo cenizo, koomági úsi, Tepehuan

Buddleia wrightii B. L. Robinson
lengua buey, Pima Bajo

Buddleia spp.
butterfly bush, English

Most butterfly bushes are native to the tropics. *Buddleia* (also spelled *Buddleja*) species are trees or shrubs with opposite leaves and small, fragrant, yellowish flowers that attract butterflies and hummingbirds. The species occurring in northwestern Mexico are commonly small shrubs, but *B. cordata* is a large tree with yellow flowers.

Buddleia

1. Used for menstrual pain, stomach disorders, wounds, bruises, burns, and respiratory and kidney conditions.

2. Has astringent, anti-inflammatory, and expectorant activity.

3. Toxicity not reported.

Historic Use. Hernández (1959, 2:281) said the Aztecs found *Buddleia* to be an effective diuretic and also used it to treat fever and wounds. The bark of the root and stem in a decoction was said to alleviate afflictions of the uterus and restore the uterus to its place. The decoction was also thought to provoke urine and clean the body.

Modern Use. The leaves and bark of *B. sessiliflora* are made into a tea and drunk for menstrual pains and stomach disorders by the Tepehuan. The leaves are crushed and moistened, then applied as a poultice to wounds. A tea made from *palo cenizo* leaves relieves discomfort caused by diarrhea. The Pima Bajo make a poultice by mixing crushed leaves with any available grease for burns. The Tarahumara warm the leaves in ashes to apply hot as a poultice or make an ointment with lard to treat wounds and bruises. The Mountain Pima apply fresh or boiled leaves to wounds.

For the Mexicans of Baja California Norte, *salvia real* is used to calm the nerves and increase energy. It helps with cold, cough, and congestion, as well as feverish respiratory tract infections and liver and gastrointestinal problems. *Buddleia crotonoides* is used in Baja California Sur. The root is used in a decoction that is drunk instead of water in appendix cases. In Arizona, Mexican Americans boil a small piece of the root of a species of *Buddleia;* the resultant beverage is refrigerated and used in place of water for kidney infection.

Phytochemistry. *Buddleia* species contain flavonoids such as rutin (Winkelman 1986:114–15) as well as aucubin, terpenes, and saponins. In one study no antibiotic activity of extract of *B. crotonoides* branches was noted (Encarnación and Keer 1991). In another, *B. americana* was shown to be effective against *Shigella flexneri* in vitro (Cáceres et al. 1993); the leaf of *B. americana* inhibited *Staphylococcus aureus* (Cáceres et al. 1991).

■ *Bursera* (Burseraceae)

ELEPHANT TREE

Bursera confusa Engelm.
totokopitkam, torote blanco, Pima Bajo; elephant tree,
   English

Bursera fagaroides (H.B.K.) Engl.
torote prieto, Pima Bajo; fragrant elephant tree, English

Bursera fragilis S. Wats
[= *Bursera lancifolia* (Schlect.) Engl.]
stukopitkam, Pima Bajo; *rusíwari,* Tarahumara; *torote prieto, torote jolopete,* Warijio; *torote papelio, torote amarillo, torote colorado,* Spanish

Bursera grandifolia Engl.
iweri, palo mulato, Tarahumara; *palo mulato,* Warijio

Bursera hindsiana (Benth.) Engl.
xoop inl (elephant tree's fingers), Seri; *copal, torote prieto,*
   Spanish; red elephant tree, English

Bursera jorullensis Engl.
copal, copalquáhuitl, árbol gumifero, Aztec

Bursera laxiflora S. Wats.
xoop caacol (large elephant tree), Seri; *torote prieto,* Yaqui,
   Spanish

Bursera microphylla A. Gray
chukuri tooro, toro chukuri, torote colorado, Mayo; *koop,*
   Seri; *copal, torote, torote blanco,* Spanish; little elephant
   tree, English

Bursera odorata Brandegee
torote blanco, Baja California Sur

Bursera penicillata (Sessé & Moc. ex DC.) Engl.
torote, Tarahumara; *torote copal, toro',* Warijio; *torote copal, torote prieto, torote incienso,* Spanish

Bursera spp.
morewá, Tarahumara; *chukuri tooro, torote prieto, toori sikiri chukam, torote colorado,* Mayo; *copal,* Spanish;
   elephant tree, English

These members of the torchwood family are aromatic trees whose gum, containing a volatile oil, exudes from wounds or broken stems. Most *Bursera* species are called *torote*, the name deriving from *toro*, the Cáhita and also Ópata name of the tree, while -*te* is the genitive form. The color *(prieto, blanco)* differentiates some species.

Historic Use. The Aztecs had various uses for *copal*, both ceremonial and medicinal. Many trees produce a gum; all are called *copalli* (resin). The resins, which varied in color depending on the species, were used in steam for 'cold' diseases, since they were classified as 'hot' and 'dry' in the third degree and astringent. *Copal* was thought to cure headache and "uterine strangulation," especially when administered in a vapor bath (Hernández 1959, 2:176), and to combat dysentery and diarrhea. Monardes (1574:fol. 3) reported that it was good for all kinds of female conditions. The seed of the fruit of another variety in ointment was considered efficacious in curing "leprosy" (Hernández 1959, 2:176–80). Gerard ([1633] 1975:1436), quoting Dioscorides and Galen, recommended frankincense for similar uses.

The Aztec uses of *Bursera* species have been continued throughout much of the American and Mexican West. In northwestern New Spain, the juice of a *Bursera* was used for fever. The resin was given in steam to induce menstruation and to treat fever. *Copal* mixed with egg white was used in a plaster to raise the fontanelle in *mollera caída* (Esteyneffer [1719] 1978:441). Fever, rheumatism, venereal disease, chilblains, rupture of the umbilicus, wounds, and fractures were all treated with *copal* in the eighteenth century.

Modern Use. The Pima Bajo make a tea from the leaves or the fruit of *B. confusa* and *B. fagaroides* to relieve coughing spells, and they place the gum on aching gums or teeth. They use gum from the bark of *B. fragilis* and *B. laxiflora* on aching gums or decayed teeth. The Yaqui cook the bark of *torote* for cough medicine. The Mayo use the heated resin of *tooro sikiri chukam* for removing thorns. The bark of *chukuri tooro* is cooked with a little cinnamon to be taken twice daily for aching limbs and for cough. The Warijio place the gum of *B. laxiflora* on an aching tooth, use the gum of *B. fragilis* as a poultice for backache, and decoct the

---

Bursera

1. Used for cough, aching teeth and gums, and fever.

2. Has antitussive, expectorant, and antibiotic activity.

3. Has toxic monoterpenes linalool, phellandrine, and terpineol.

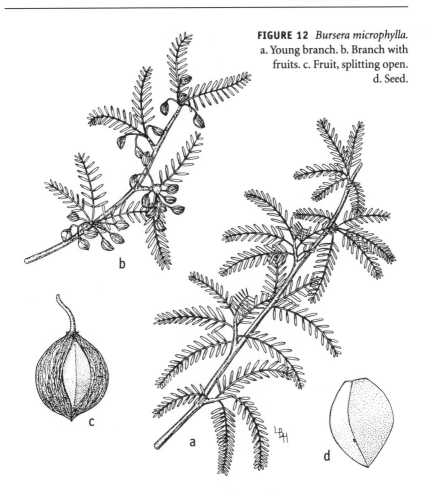

FIGURE 12 *Bursera microphylla.*
a. Young branch. b. Branch with
fruits. c. Fruit, splitting open.
d. Seed.

bark of *B. grandifolia* for fever, especially malaria. They use
the herbage and bark of *B. penicillata* for a cold. The Tarahumara make a hot infusion of the bark of *B. grandifolia*,
which they drink for body pain. They use the resins of several Burseraceae to treat pain in the ear and eye, to relieve
the pain of burns, and for general curing.

The Seri take shavings from *xoop* wood and cook these
in water with leafy branch tips of *Hyptis* for a tea to cure
difficulty in breathing for asthmatics. They cook the leaves
in water with *Atriplex barclayana* to bathe the bite of a
stingray and make a decoction of the leaves with *Stegno-*
sperma as a shampoo to cure a headache. They apply *xoop*
leaves to help dry the umbilical cord stump. A shampoo
made with its fruit is used to kill head lice, and a tea made
from the bark is said to cure gonorrhea. The cortex of the

tree is mashed with water into a paste and applied to sores on the head. Strips or pieces of *torote prieto, xoop caacol,* bark are used to make a tea for cold and cough. The dark portion of the bark is used for a tea taken to calm the pain of a scorpion or black widow bite.

The Mexicans in Baja California Sur make a decoction from the branches of *B. microphylla* to wash wounds and bruises. The gum of *B. odorata* is rubbed on scorpion stings. A Tucson *curandero* makes a tea by boiling the bark: he recommends taking one cup in the morning and another in the evening to fortify the body against allergies. The resin can be purchased in Tucson.

Phytochemistry. According to NAPRALERT, some *Bursera* species contain active compounds, including triterpenes, monoterpenes, flavonoids, and steroids. *Bursera fagaroides* contains beta-peltatin A, and *B. microphylla* has podophyllotoxin. Linalool is present in *B. penicillata*. The dried bark of *Bursera simaruba* shows antifungal, cytotoxic, and diuretic activity. The resins contain aromatic triterpenes and etherial oils (Johnson 1992). An extract of branches of *B. microphylla* proved active against *Staphylococcus aureus* and *Bacillus subtilis*, and *B. odorata* is active against *Staphylococcus aureus, Bacillus subtilis*, and *Streptococcus fecalis* (Encarnación and Keer 1991).

■ *Caesalpinia* (Fabaceae)
MEXICAN BIRD-OF-PARADISE

Caesalpinia mexicana A. Gray
juuya tabachín, tabachín del monte, Mayo

Caesalpinia platyloba S. Wats.
palo colorado, Tarahumara

Caesalpinia pulcherrima (L.) Swartz
chamolxóchitl, cacalaca, Aztec; *tabachín,* Ópata; *tavachín,* Pima Bajo; *makapal,* Tarahumara; *talpakachi,* Warijio; *noche buena, tabachín,* Spanish; Mexican red bird-of-paradise, English

Bird-of-paradise is a small tree with numerous small leaflets and showy yellow and red flowers. The Aztec name of the species came from its similarity to red feathers; the Latin genus was named for a sixteenth-century Italian botanist; the species name, *pulcherrima*, translates as "most beautiful."

Historic Use. Hernández (1959, 2:309) said the Aztecs believed that the flowers were 'dry' and astringent and cured tumors. The pods after being dissolved in water and instilled in the nose stopped bleeding. According to Nuñez (1777), the Ópata found the seed of *el tabachín* to be poisonous but the root, very medicinal. A medicine to stop diarrhea was prepared as follows: cut the root in the evening near sunset, scrape the upper side of the root, chew or grind it, cook it in a little water, and drink it at dawn.

Modern Use. The Tarahumara apply the bark and make a wash of the leaves of *C. platyloba* for toothache. They use tea made from the root of *C. pulcherrima* for fever and for venereal disease. The Mayo cook the pods of *C. mexicana* with half a tablespoon of salt and take one cup twice daily for stomach infections. The Warijio boil the roots of *C. pulcherrima* in water for treating insect stings and snakebites. The Mountain Pima make a lotion of the roots and strip branches of a *Caesalpinia* species that they apply to wounds. The Pima Bajo dry the roots of *C. pulcherrima*, grind them, and place them on the bite or sting of an insect. They crush and boil, then drain the roots for a poultice on wounds. They also use the roots in a gargle for throat disorders.

No peoples in the American and Mexican West are reported to use *C. pulcherrima* bark or flowers as an emmenagogue or to induce abortion, although these are common applications throughout the tropical world. Such use also was not reported by Mexican Americans in Tucson, who apparently purchase the herb for some purposes at certain supermarkets.

Phytochemistry. As cited in NAPRALERT, sixty-three studies of *C. pulcherrima* have been conducted throughout the world. Most found *C. pulcherrima* inactive against bacteria (except for dried bark against *Staphylococcus aureus*, in Tanzania) and viruses but found strong antifungal action. Tannins, diterpenes, coumarins, and flavonoids were found in the bark. A protein mixture from the seed demonstrated

Caesalpinia

1. Used for snakebite and insect stings, toothache, fever, and venereal disease.

2. Has tannins and shows antifungal activity.

3. Toxins include hydrocyanic acid, phellandrine, and shikimic acid.

good potential antitumor activity. Toxins of *Caesalpinia* include hydrocyanic acid, phellandrene, and shikimic acid (Duke 1985:550).

■ *Cannabis* (Moraceae)

MARIJUANA

Cannabis sativa L.

> hempe, Renaissance English; *cañamo*, northwestern New Spain; *'o 'oB yai paim*, Paipai; *marijuana*, Spanish; marijuana, marihuana, hemp, pot, English

Marijuana is a major illicit cash crop in northwestern Mexico. It is an herbaceous plant with leaves shaped like a hand; its strong, fibrous stem is used for making cloth and rope.

Historic Use. *Cannabis* is an Old World remedy. As far back as 6000 B.C. it was esteemed in China (Lewis and Elvin-Lewis 1977:429). Galen said that *Cannabis* was so great a drier that it dried up the "seed" if too much was eaten of it, and Dioscorides added that it assuaged earache if the juice was dropped into the ear (Gerard [1633] 1975: 708–9). In the eighteenth century, Segesser (n.d.) included a request for *Cannabis* seed when writing to his brother in Switzerland for herbal remedies. Esteyneffer ([1719] 1978: 447) recommended eating the seed of *cañamo* to decrease mother's milk. To prevent seminal emission (called *purgación*), he said, it should be taken in a drink made of barley water and sugar (Esteyneffer [1719] 1978:419).

Modern Use. Today I am told that marijuana is a seldom used anodyne for rheumatism. The Mayo mix the leaves with alcohol and apply to the aching place. The Paipai use it for colic. Mexican Americans soothe the pain of rheumatism by rubbing the affected part with a lotion made by soaking the leaves of marijuana with an avocado pit in alcohol.

Phytochemistry. Delta-tetrahydrocannabinol is the principle hallucinogen (Duke 1985:96–97). *Cannabis* may interrupt DNA synthesis and cause chromosomal damage. Constituents of seeds that might relate to pharmacological

---

Cannabis

1. Used externally in liniment for rheumatism.

2. Is hallucinogenic and lipotropic.

3. Pollen is allergenic. Ingestion may interrupt DNA synthesis.

effects are choline (lipotropic), inositol (a growth factor), phytosterols, and trigonelline. The pollen is an allergen.

■ *Capsicum* (Solanaceae)

CHILI

Capsicum annuum L.
   ginnie pepper, Renaissance English; *chilli,* Aztec; *korí,*
      Tarahumara; *chili,* Spanish; chili, chili pepper, English

Chilis are small herbaceous plants bearing dry, strong-tasting fruits that turn red as they mature. Wild chiltepine fruits are the size and shape of peas.

Historic Use. The botanical name *Capsicum* comes from the Greek word meaning "to bite." Columbus and the official physician of his second voyage, Diego Alvares Chanca, were impressed by chili and took it back to Spain. Hernández (1959, 2:136–39) described and had several varieties illustrated, noting the plant's resemblance to *Solanum.* He decreed *chilli* to be 'hot' in the fourth degree and 'dry' in the third. He said that it provoked urine and menstruation, and strengthened a stomach debilitated by 'cold'; in salsa it excited the appetite. It purged the body of all "pititous" humors (which were 'cold' and 'wet'), especially in the hip joint. If used immoderately it could irritate the kidneys, inflame the blood and liver, or cause disease of the kidneys, brain, and pleura, leading to peripneumonia and other internal inflammations and eruptions. Gerard ([1633] 1975: 364–66) similarly described and illustrated different varieties, also noting the plant's resemblance to nightshade *(Solanum).* He, however, believed it had little to recommend it medicinally except that it dissolved swellings about the throat.

   In northwestern New Spain, Esteyneffer ([1719] 1978: 160, 161, 295) recommended chili to balance an excess of "pituita" (the 'cold', 'moist' humor), thus curing headache. In a salsa, it stimulated the appetite. However, he said that it should be avoided by those with liver trouble.

Modern Use. The Pima Bajo give a *chiltepín,* the smallest chili, as medicine to people who develop stomach disorders

---

Capsicum

1. Used against pain.

2. Capsaicin stimulates salivation and sweating, reduces blood coagulants and triglycerides, and strengthens heartbeat.

3. Is high in vitamin C. Can produce dermatitis and irritate mucous membranes.

from working in the smoke and dust of the mines. The Mayo mix the leaves with alcohol to make a liniment for rheumatism. The Tarahumara chew the fruit of *Capsicum* with other plants for headache. Mexican Americans dip a cotton swab in hot red chili, then touch it to tonsils to relieve pain. They treat earache by putting ground chili in the ear, holding it in place with cotton. It is sold as a medicinal.

Phytochemistry. The hot principle in the fruit, capsaicin, is still noticeable at 1:11,000,000 dilution (Duke 1985: 98–99). Capsaicin stimulates salivation and sweating. The plant contains solanidine, solanine, solasodine, and scopoletin. Some research has suggested that capsaicin may contribute to liver cancer. Studies (see Rodriguez 1992:4) have shown that capsaicin is also an antioxidant, is high in vitamin C, and retards the production of nitrosamines. It reduces blood coagulants, reduces arterial blood pressure, strengthens heartbeats, and reduces blood triglycerides. Chili can produce dermatitis and is an irritant to mucous membranes.

■ *Carnegiea* (Cactaceae)

SAGUARO

Carnegiea gigantea (Engelm.) Britt. & Rose
has shan, Pima; *pitahaya* (fruit), Mayo; *mojépe, sahuaro,*
Seri; sahuaro, saguaro, Spanish and English

Carnegiea

1. Has been used against rheumatic and hemorrhoid pain and in gruel to make milk flow.

2. Contains isoquinoline alkaloids and dopamine, capable of hallucinatory action.

3. No reports of use in hallucinatory rites.

Saguaros are huge columnar cacti that are found only in southern Arizona and Sonora. They frequently develop lateral "arms" when mature. They bear white flowers at the very tops of the cactus and its arms; the flower becomes a delicious red fruit used in making jam, syrup, and wine. The fruits of several of the large Cactaceae are commonly called *pitahaya.*

Historic Use. In northwestern New Spain, Esteyneffer ([1719] 1978:552) noted that the *pitahaya* fruit was an important part of the diet to cure a *flegmón,* or cellulitis (that is, an inflammation of cellular tissue), for which he gave the classical description: "*tumor . . . con calor, rubor, pulsación y tensión,*" a swelling with heat, pain, redness, throbbing, and tension.

Modern Use. The Pima make a gruel by grinding the seeds on a metate, mixing the ground seed with an equal quantity of whole wheat, and adding boiling water. This gruel is given to women to "keep the stomach warm" and make milk flow after childbirth. The Mayo prepare *pitahaya* for treating hemorrhoids by drying, grinding, and cooking the flesh of the fruit in cooking oil. A bandage is dipped in this mixture and applied to the hemorrhoid. The Seri use the stem for rheumatism: the spines are removed from a slab, heated in hot coals, wrapped in cloth, and placed on the aching part. I have no report of Mexican Americans using saguaro for healing purposes today.

Phytochemistry. According to NAPRALERT, phytochemical studies of *C. gigantea* have found isoquinoline alkaloids labeled arizonine, carnegine, and gigantine, as well as dopamine, heliamine, salsolidine triterpenes, and alicyclic compounds in the entire plant and steroids in the pollen. Seed analysis is not reported.

■ *Casimiroa* (Rutaceae)

ZAPOTE

Casimiroa edulis S. Wats.
cochizizáitk, cochitzapotl, tzápotl somnifero, Aztec; *zapote, zapote blanco,* Spanish

Zapote is a large tropical tree with edible yellow fruits up to ten centimeters (four inches) in diameter. Native to Mexico, it has no English common name.

Historic Use. The Aztec name for the medicinal translates as "sleep-inducing sweet fruit." Hernández (1959, 2:92) noted that the *zapote* looked like our apple, but he said, "these are not healthy to eat and the seed is poisonous." It was eaten to help one sleep; ancient Aztec theory located the soul, mind, and consciousness in the heart (Lozoya 1980:89). The burned and pulverized seed was used to cure a putrid ulcer: consuming bad tissue, cleaning the ulcer, growing new flesh, and producing cicatrization rapidly. The leaves were mashed and applied to the breast of a wet nurse to cure an infant of diarrhea.

Casimiroa

1. Used for heart problems and high blood pressure.

2. Has hypotensive and sedative actions.

3. No toxicity reported.

In northwestern New Spain, Esteyneffer ([1719] 1978: 747) recommended a little of the pulverized seed of *zapote* to evacuate disease-causing phlegm from the head, stomach, and other parts. Clavijero reported the seed to be poisonous, the fruit, sleep inducing.

Modern Use. *Zapote* leaves are available in pharmacies of the San Luis Valley in Colorado and are sold in Mexican herb stores for treating the heart. *Zapote* is also taken by Mexican American patients for problems with high blood pressure; in fact, one anxious daughter from Sonora brought it to treat her father while he was a patient in the University of Arizona Medical Center Hospital. It may be purchased at a Tucson *yerberia* and certain supermarkets.

Phytochemistry. *Casimiroa edulis* contains N,N-dimethylhistamine, which has antihistamine action and thus acts as a sedative, and casimiroin and fagarine, which have hypotensive and sedative actions (Lozoya and Lozoya 1982:130–73). In animal studies, the alkaloid alocriptopine is valued for atrial fibrillation, while the alkaloids protopine, berberine, and sanguinarine stimulate the tone and strength of cardiac contraction, prolonging the refractory period. Berberine also has antibiotic properties; in addition, oral administration results in a lowering of bilirubin levels while increasing bile. In Guatemala, where *C. edulis* is used for gastrointestinal diseases, it has been found effective against *Shigella flexneri* (Cáceres et al. 1993).

■ *Cassia* (Fabaceae)
SENNA

Cassia confinis E. Greene
ojasen, Baja California Sur, Baja California Norte

Cassia covesii A. Gray
hehe quiinla (plant that rings), Seri; *hojasen*, Pima Bajo, Spanish; desert senna, English

Cassia grandis L.
cuarto quauhayohuachtli, cañafistula, Aztec; *cañafistula*, northwestern New Spain, Spanish; pudding pipe tree, senna, common English

Cassia leptadenia Greenm.
hojasen, Tepehuan

Cassia occidentalis L.
tercer tlalhoaxin, ecapatli, medicina del viento, Aztec

Sennas are trees and shrubs with bright yellow flowers that may be found widespread in tropical climates.

Historic Use. The generic name *Cassia* was the Greek name for a species of this genus or a related genus. In the Old World, Gerard ([1633] 1975:1431) reported that the pulp of *Cassia fistula* could be given without danger to all weak people, "for it gently purgeth cholerick humours and slimie phlegm"; he recommended it for agues, pleurisies, and jaundice or any other inflammation of the liver, as well as for the kidneys and reins (loins) and other conditions, assigning to it the properties of 'moist' and 'hot'.

The Aztecs had seventeen plants that might have been *Cassia* species. They used several species of *Cassia* for inflammation and as a purgative. The plant now identified as *C. grandis* "evacuates *bilis* and phlegmatic humors," judged by Hernández (1959, 2:56) to be 'hot', 'dry' and a little astringent and said to cure tumors and ulcers. The leaves of another species, now identified as *C. occidentalis,* when mashed and placed on the stomach helped infants who vomited their milk, and when applied as an ointment relieved headache and chills of fever (also *empacho* and leprosy). It was grown everywhere as a remedy (Hernández 1959, 2:131). In northwestern New Spain, *cañafistula* was recommended for retained menstruation, hemorrhoids, and gonorrhea, conditions that were believed to require a purgative (Esteyneffer [1719] 1978:814).

Modern Use. The Pima Bajo take a tea of the roots of *C. covesii* as a childbirth aid. A tea of the flowers is used as a purgative. The Tepehuan boil the whole plant of *C. leptadenia* for a few minutes for a fever treatment. The Mountain Pima (no common name given) find that a drink from the flowers of *C. occidentalis* gives energy to the elderly. In Baja California Sur, a decoction of *ojasen* leaves is used as a purgative to cure stomach troubles. The decoction is also recommended in Baja California Norte. The Seri drink a tea from the roots of *C. covesii* to stimulate the appetite and

Cassia

1. Used as a purgative, as a childbirth aid, and for fever.

2. Some species have laxative and estrogenic compounds.

3. Leaves may cause severe dermatitis. Ingestion may cause cramps and colon problems.

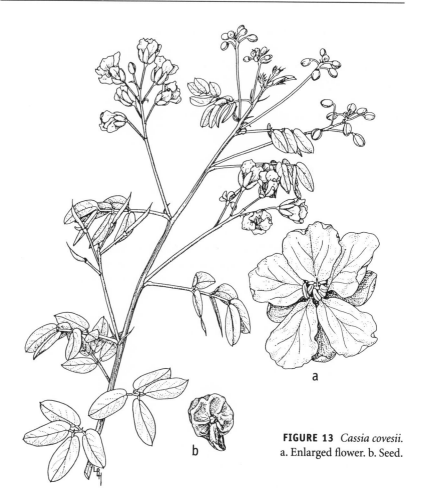

FIGURE 13 *Cassia covesii.*
a. Enlarged flower. b. Seed.

clean out the stomach. This tea is also used to treat kidney problems and measles and helps to bring about conception. A tea of the leaves and the stems is taken for the liver and for chicken pox. Mexican Americans buy *cañafistula* to use as a purgative.

Phytochemistry. The seedpods or stem bark of *Cassia fistula,* an Indian species, showed antiviral, cytotoxic, and hypoglycemic effects in vitro (Tewari 1979). Other species of *Cassia* that are sold as laxatives contain sennosides, the laxative principles. Purgatives have been considered emically to be appropriate treatment for fevers and stomach and kidney problems. Compounds that might explain use in terms of biomedical theory include salicylic acid, chrysophanic acid, anthraquinones, beta-sitosterol, and saponins (Duke 1985:102). An extract of *C. confinis* branches was found to have 2+ activity against *Staphylococcus aureus* (Encarnación and Keer 1991). In Argentina, an extract made with boiling water of *C. occidentalis* root was found to be effective against *Salmonella typhi* (Pérez and Anesini 1994).

■ *Cereus* (Cactaceae)

CACTUS

Echinocereus spp.
quacuetzpalcuitlapilli, cola de iguana, Aztec; napísala, Tarahumara; pitayita, Spanish; hedgehog cactus, English

Lophocereus schottii Britt. & Rose
musue, muso, savila etcho, Mayo; garambullo, Baja California; senita, pitayita, garambullo, Spanish; old man cactus, English

Pachycereus pecten-aboriginum (Engelm.) Britt. & Rose
wichowaka, Tarahumara; echo, etcho, pitahaya, Spanish; hairbrush cactus, native's comb, English

Pachycereus pringlei (S. Wats.) Britt. & Rose
echo, hecho, northwestern New Spain, Mayo; cardón, cardón hecho, Cochímies; wichowaka, Tarahumara; xaasj, cardón, Seri; sahueso, cardón, Spanish

Peniocereus greggii (Engelm.) Britt. & Rose
ho-ok vaao, Pima; reina de la noche, saramatraca, Spanish; night blooming cereus, English

Stenocereus thurberi (Engelm.) F. Buxb.
[= *Lemaireocereus thurberi* (Engelm.) Britt. & Rose]
pitahaya tepexinse, Aztec; aki, Mayo; ool, Seri; pitaya dulce, Spanish; organ pipe cactus, English

These genera are grouped together and were formerly subsumed in one genus, *Cereus* (Kearney and Peebles 1942). *Lophocereus, Pachycereus, Peniocereus,* and *Stenocereus* species are columnar cacti that acquired the -*cereus* name from the Latin word meaning "long candle" and their generic name from their appearance. *Echinos,* Greek for hedgehog, is the prefix name for a low cactus covered with straight spines. *Pachy* refers to the thick stem of the *cardón*, which is the world's largest cactus. It is similar to the saguaro but larger and more profusely branched. The species name *pecten-aboriginum* is Latin for "native's comb." *Lophos* (a crest) and *phoreo* (to bear) refer to the tufts of woolly hairs on the upper portions of the old-man cactus. The common name *pitahaya* is used for these large branched columnar cacti, and it is also applied to the fruit. Night blooming cereus does bloom at night. It has trailing stems and extremely fragrant white flowers.

Cereus

1. Used for earache, wounds, and pain.

2. Psychoactive, cardiotonic actions.

3. Potential for cardiac toxicity and gastrointestinal upset.

Historic Use. The Aztecs made much medicinal use of these plants. Hernández (1959, 2:237) said they employed one for fever and to alleviate erysipelas and any other inflammations. From the text and drawing illustrating this cactus, it appears to be an *Echinocereus*. Hernández (1959, 3:102) heard that the root of another cactus (now believed to be a *Stenocereus*) cures swelling of the belly and hydrops (dropsy), but he did not know how it could do that without some special property since, according to his system, he categorized the cactus as 'cold' in the second degree and 'moist', the same properties as the condition to be cured.

Eighteenth-century reports from northwestern New Spain all recommended *pitahaya* or *hecho*. The Ópata gave the fruit, cooked or raw, for complaints of the stomach. Pfefferkorn ([1794–95] 1989:76) said the fruit was a powerful antiscorbutic. In 1783, Amarillas recommended a cure of the Yaqui that he said was very painful but healing for wounds: making a balsam of the juice and soaking dressings in this for application. Pablos [1784] at Rio Chico used the bark instead, while Tamayo [1784], stationed at Arivechi, said to cook a piece of the trunk and gargle with the cooking water for ulcers in the throat. Clavijero ([1786] 1937:37) reported that the Cochimíes in Baja California made a balsam for wounds and sores by squeezing the juice out of a piece of *cardón*, pounding and boiling the piece, then removing the scum.

Modern Use. The Pima make a concoction of *reina de la noche* by cutting up and boiling the large, tuberous root, then taking the concoction to cure diabetes; or the tubers may be sliced and the juice sucked out. They cure venereal disease with a tea made from the bark; they drink this tea in the early morning hours for about one week. The Tohono O'odham mix the seedpod of a *Cereus* (no common name given) with grease to make an ointment, which they rub on sores. The Tarahumara use the stem of *Pachycereus pringlei* in a tea for general medicine and pain as well as a purgative. They also use it as a hallucinogen.

The Mayo make extensive use of these cacti. They use *Pachycereus pecten-aboriginum* for healing. They cut a piece of trunk and anoint a wound with the cut surface to stop hemorrhage; for infected wounds, some of the trunk is cooked in water with a little salt, and the infected part is washed with this solution three times daily, followed by sulfathiazole powder. They press juice from the stem of *etcho* or cook it in water, which they drink when cool, to treat a sore throat. For the bites of black ants, a small piece of *Stenocereus thurberi* is cut and the stung part is anointed with the juice. The pith is cooked and drunk daily like water to serve as a purgative. They use *Lophocereus schottii* to cure cancer. The whole plant is cooked in water to make a daily drink. For ruptured eardrums, a piece is boiled and the juice is squeezed out, then dropped into the ears. The Mayo take a piece of *muso*, boil it with water, and let it cool. They drink this daily for kidney problems and body aches and to improve circulation. They claim that there are two kinds of *muso*, one with five ridges and one with six ridges: the latter is medicinal.

The Seri take fresh slabs of *Stenocereus* with the spines removed, heat them in coals, wrap them in cloth, and place them on aching parts of the body. They wrap the stillborn infant or miscarriage in a piece of cloth, place it in a box, and then put it in the limbs of the *cardón;* the placenta is buried at the base of the *cardón,* and ashes are put on the top of the burial to keep coyotes away.

In Baja California, sliced *Lophocereus schottii (garam-bullo)* is sold in markets as a treatment for stingray wounds. A tea is made to help relieve ulcerated stomachs and treat ulcers (Roberts 1989:131). Mexicans in the sierras of Sinaloa use the juice of *Pachycereus pecten-aboriginum* as a mouth-wash to check bleeding, especially following tooth extraction (Werner 1970:65). A Tucson *curandero* tells me that every kind of cactus is very medicinal, and he grows several in his garden. I have not found any of these cacti in *yerbe-rias*, however.

Phytochemistry. Little information is available on the pharmaceutical actions of Cactaceae. Since many peoples use these plants for pain, one would expect it to contain alkaloids that give relief. One study does report that the stem and flowers of *Selenicereus grandiflorus* Britt. & Rose (called night blooming cereus) contain the alkaloid cactine, which has activity similar to that of digitalis and may be the same as hordenine; fresh juice burns the mouth and causes vomiting and diarrhea (Duke 1985:441). The flowers contain rutin, narcissin, cactacine, kaempferitin, grandiflorine, and rutinosides. The root contains peniocerol, viperidone, desoxyviperidone, viperidinone, beta-sitosterol, and (probably) caffeine (Moore 1989:80). According to NAPRALERT, studies have been conducted to identify psychoactive principles in *Pachycereus*. Several isoquinoline alkaloids and quinic acid were found. Four *Cereus* species (*C. donkelaarii, pitahaya; C. rostratus, tuna de tlacuache; C. senilis, cabeza de viejo;* and *C. serpentinus, organillo*) are listed as having cardiotonic actions (Lozoya 1980).

■ *Chenopodium* (Chenopodiaceae)
GOOSEFOOT, WORMSEED

Chenopodium ambrosioides L.
epázotl, hierba olorosa, Aztec; *patusai*, Tepehuan; *basote,* Tarahumara; *epazote, ipasote,* Pima Bajo, Mountain Pima, Spanish; goosefoot, Mexican tea, Spanish tea, wormseed, English

Chenopodium botrys L.
botrys, ambrosia, oke of Cappadocia, Renaissance English; Jerusalem-oak, feather-geranium, English

Chenopodium graveolens Willd.
la sadugua, Ópata; *chu'ja,* Tarahumara; *yerba del zorrillo, yerba del chivatito,* eighteenth-century Spanish; *epasote,* Spanish; goosefoot, English

Chenopodium quinoa Willd.
quinoa, Spanish, English

Chenopodium spp.
epasote, New Mexico Spanish, Colorado Spanish

Epasote is a common strongly scented herbaceous plant with tiny yellow flowers. *Chenopodium* gets its botanical name from the Greek words for goose and foot, referring to the shape of the leaves.

Historic Use. Dioscorides said that botrys (Old World *Chenopodium*) was mainly used to place in clothing because its odor repelled clothes moths, but boiled in wine it also helped breathing (Font Quer 1979:153). *Chenopodium botrys* has been extensively naturalized in America.

    Chenopodium ambrosioides is native to the New World, but it was exported to Spain in the eighteenth century. There it has invaded and is called *té de Nueva España* or *té de Méjico*. The earliest evidence of possible medicinal use in the American and Mexican West is from analyses of coprolites (Reinhard, Ambler, and McGuffie 1985). A well-known vermifuge in biomedicine today, its use for that purpose was earlier recorded by Sahagún and Hernández. The Aztecs used it raw or cooked and mixed with food to strengthen and alleviate those with asthma or other chest conditions. They employed it to "expell harmful animals from the belly and extinguish inflammation." Hernández (1959, 2:369) said it was "sharp, fragrant and 'hot' in the 3rd degree." This is one of the most widely used plants in Mexico (Lozoya and Lozoya 1982).

    In northwestern New Spain, Esteyneffer ([1719] 1978) recommended making a decoction from powdered leaves. This decoction was drunk for a multitude of conditions, including rheumatism, pituita disease, sincope, *tabardillo* (typhus), and burns. Nuñez (1777) said that the name given by the Spaniards, *hierva (yerba) del zorillo* (skunk herb), may come from the excessive odor of the root (the second name of the binomial, *graveolens*, also reflects the odor). It was good for all pains, especially those that came from 'cold', because it was a very 'hot' herb.

Modern Use. The Tepehuan make a tea from the leaves of *C. ambrosioides* that is given to facilitate childbirth and to expel intestinal worms. The Mountain Pima boil the whole plant for the recently delivered. The roots make a drink for

Chenopodium

1. Given to expel worms, for gastrointestinal problems, and to facilitate childbirth.

2. Has a strong purgative action.

3. Therapeutic dose is close to poisonous level.

colic. The Pima Bajo boil the whole plant to make a decoction for stomach disorders. The Tarahumara drink the tea of *C. ambrosioides* or *C. graveolens* to alleviate stomach problems. They consider *C. graveolens* good for diarrhea. For a vermifuge, the aboveground parts of these plants are used in a tea, which is drunk after the main meal. It also is taken to relieve stomachache, headache, and fever. The plants are also used as a condiment but are known to be dangerous if not gathered before the fruit matures. The Navajo and Zuni also use *Chenopodium* species.

Mexicans buy *epazote* leaves in herb stores to make a tea to expel parasites and treat stomachache. Mexican Americans in southern New Mexico brew a tea of the leaves, which they give to an infant to prevent colic. If the breast-feeding mother is eating highly gas-forming foods, she drinks the tea. The same tea is good for bronchitis; for chronic bronchitis the tea should be made together with an unidentified plant *higado de zorillo* and taken four times daily for fifteen days. A tea of the leaves brings on menses, helps menstrual cramps, and treats diarrhea. An infusion of a *Chenopodium* is used as a bath in San Luis Colorado to treat fevers and in scarlet fever to prevent the hair from falling out. The plant is also used in a tea to treat "cold in the stomach," for the heart, and for children who wet the bed. Women also douche with the leaves of *epasote* as an emmenagogue, an abortifacient, and a galactogogue and to relieve postpartum pain. Quinoa is sold in an herb store in Sonora. A decoction of the seed is used as a wash to treat sores.

Phytochemistry. The oil of *C. ambrosioides* is dangerous (Duke 1985:114). It contains ascaridole, p-cymene, camphor, and 1-limonene, along with numerous sesquiterpenes and flavonoids. The therapeutic dose is close to minimum poisonous levels, causing fatalities. Laxatives are considered appropriate treatment for fever and for various stomach problems in various folk systems. According to NAPRALERT, in a study of *C. botrys* conducted in Saudi Arabia, antibacterial activity was found against *Bacillus subtilis* and *Streptococcus aureus*.

■ *Citrus* (Rutaceae)

LEMON, LIME, ORANGE

Citrus aurantifolia Swingle
lima, Spanish; lime, English

Citrus aurantium L.
naranja agria, northwestern New Spain; *naranjo* (tree),
   naranja (fruit), *azahar, flor de azar* (orange flower),
   Spanish; orange, English

Citrus limon Burm.
limón, Spanish; lemon, English

Citrus trees are commonly planted Old World trees with
sharp spines and sour fruits.

Historic Use. In the Old World Gerard ([1633] 1975:1462–
65) discussed all the citrus together, noting that the juice
was 'cold' and 'dry', while the rind and seed were 'hot' and
'dry'. Quoting Galen, he recommended citron for the appetite, for fever, and against all pestilent and venomous or
infectious diseases. He said it cured mange and itch and "it
comforted the heart, cooled the inward parts, and thinned
gross, tough, and slimy humors."

Citrus trees were introduced to Mexico, where they
soon became known for medicinal uses. Orange was a popular remedy in northwestern New Spain, recommended for
all conditions caused by 'cold', especially melancholy (the
'cold' and 'dry' humor). Esteyneffer in 1712 reported that
lime juice was good for *escorbuto*, or *mal de loanda*, as he
called it: scurvy. (Loanda or Luanda is the capitol of An-
gola, which was a Portuguese settlement at that time.) He
said that the disease had not long been known, but his description of the pathology of scurvy could stand today. He
believed that it was caused by melancholic humors. To treat
scurvy, he recommended watercress and parsley or sorrel,
or a Sonoran species of cress, *oyvari*, and a Mexican sorrel,
xoxocoyolli (note that these have a high vitamin C content).
If lime juice was not available, radish juice might be used.

Gerard's differential categorization of the juice as 'cold'
and the rind as 'hot' was followed by Esteyneffer ([1719]
1978:161), who recommended conserve of orange or lemon

Citrus

1. Tea of orange blossoms
is used to treat nerves and
heart; hot lemon tea is
taken for colds.

2. May have diuretic, ex-
pectorant, and antibiotic
activity.

3. No reports of toxicity.

rind for headache (a 'hot' disease). A decoction of orange rind or rind powdered in an *atole* (corn gruel) was good for *el saguaydodo*, or yellow vomit of Sonora (Esteyneffer [1719] 1978:304). Och (Treutlein 1965:171) reported the same cure for the disease. The powder was an ingredient for colic caused by 'cold'. He also chose the juice for 'cold' diseases such as melancholy. The stem could be put in the mouth of a comatose person to keep it open, and the cut fruit could be placed at the nostrils to revive him. The juice was good for persons suffering *tabardillos* (typhus) and other pestilent fevers. It was also good to depress the appetite. Although Och recommended citrus fruits for many conditions, he said that eating unripe fruit, especially lemon or orange, caused retention of menstruation. However, the sour green oranges when prepared in a special drink and taken in the morning could stanch excessive flow, and the same was good for erysipelas. To cure herpes and mange, the sour orange was mixed with sugar and applied. All these skin conditions were considered to be caused by a 'hot' humor. Pfefferkorn ([1794–95] 1989:218) said that *limón* mixed with sugar and water successfully treated victims of a terrible epidemic brought with the wind from the southern part of New Spain in June 1765. The contagion, he said, started with a terrible headache and body pains, fever, and delirium, followed by vomiting and death. Blood dripping from the nose was a sign of recovery. Gilij (1785) recommended taking a potion made by mixing two ounces of orange juice mixed with two of sugar and two of wine for three or four mornings to cure tertian fevers.

Modern Use. Orange blossom tea is one of the most commonly used remedies in the American and Mexican West (as elsewhere since ancient times). The Mayo use the blossoms of any citrus for the nerves and for despair. Mexicans consider it good for the nerves and the heart. They prepare the leaves in a tea for stomachache. The Tarahumara make a tea of the leaves of *Citrus* species for headache and colds. In Baja California Sur, a decoction of the flower or bark is drunk for the heart and also for nerves.

Mexican Americans make extensive use of *Citrus*. They brew a tea of orange flowers, *manzanilla, yerba buena,* and rose petals to drink in the morning instead of coffee, and also at night, for menstrual pain. For nervousness they make a tea by brewing orange rinds. A tea of rose leaves and cinnamon sticks may also be added. Orange blossom tea is recommended for the heart and for *alta presión* (high blood pressure). For colds, they make a hot tea of lemon juice and honey. For fever, lemon tea made with whiskey and cinnamon is taken every four hours. To treat a foreign body in the eye, one drop of lemon juice is considered effective but painful. Formerly, traditional Mexican American midwives in Tucson would drop lemon juice in the eyes of the newly delivered babies to prevent disease. In Tucson *té de azahar,* orange blossom tea, is used for *nervios. Azahar* is one of the most popular herbs sold in a Tucson *botanica* and in grocery stores, although many women collect their own by placing sheets under orange trees when they are in bloom.

Phytochemistry. The printout from NAPRALERT was one and a half inches high: *Citrus* species have been subjected to numerous studies. *Citrus aurantifolia* was found to be an active diuretic. Some studies reported that the essential oil was active against *Pseudomonas aeruginosa, Staphylococcus aureus, Escherichia coli, Candida albicans, Kloeckera apiculata, Salmonella typhimurium, Aspergillus* species, *Cladosporium herbarum,* and *Penicillium cyclopium. Citrus* acts as a coagulant and central nervous system depressant. Fewer activity effects were reported for *C. limon,* which showed activity against *E. coli, P. aeruginosa, S. aureus, Salmonella typhi, Serratia marcescens,* and *Shigella flexneri,* many of the same microorganisms.

Among the compounds found in *C. aurantifolia* fruit and peel are cineol, coumarin (3.9% in the essential oil), alpha- and beta-pinenes, and terpineol. These as well as hesperidin have been found in *C. aurantium.* The flower contains linalool and 31 mcg/g caffeine. *Citrus limon* contains beta-bisabolol, coumarins, and other compounds including limonene, liminoids, hesperidin, tyramine, and stachydrine. *Citrus* fruits contain carotenoids and are well known for their vitamin C content.

■ *Datura* (Solanaceae)

JIMSON WEED

Datura discolor Bernh.
kodop, Pima; *uchiri*, Tarahumara; *toloache*, Baja California Sur; *hehe camóstim* (plant that causes grimacing [from being crazy]), Seri

Datura lanosa A. S. Barclay ex R. A. Bye
'hákundum, Pima; *rikúhuri, tukúwari*, Tarahumara

Datura metel L.
toloache, taguaro, Ópata

Datura meteloides DC. ex Dun.
[= *Datura inoxia* Mill.]
tocorhobi, hierba ponzoñosa, gugudua'gcama, Pima Bajo; *dekúba*, Tarahumara; *shmalktuch*, Paipai

Datura quercifolia H.B.K.
toloache, Mountain Pima

Datura stramonium L.
stramonium spinosum, thorny apple of Peru, Renaissance English; *tlápatl*, Aztec; *tlapa, piñon de Tlapa*, northwestern New Spain; *tokoraki*, Tepehuan; *shmalk tuch*, Paipai; jimson weed, poisonous nightshade, English

Datura wrightii Regel.
tecuyawi (*tekú*, to be drunk), *toloachi*, Warijio

Datura spp.
toloache, Mayo

Jimson weed is a dark green herbaceous plant with huge trumpet-shaped flowers up to twenty centimeters (eight inches) long, its beauty made famous in the paintings of Georgia O'Keefe. *Toloache* is the most frequently encountered common name for *D. lanosa*, which perhaps is the most common species of those labeled *toloache* in northwestern Mexico. "The native names applied by ethnic groups appear to be based upon the deliriant effects produced by the plant on the nervous system" (Bye, Mata, and Pimentel 1991:32–34).

Historic Use. In the Old World, Gerard ([1633] 1975:348–49) reported on *stramonium spinosum*, or thorny apple of

Peru, and placed it among the nightshades, saying that the whole plant is 'cold' in the fourth degree and of a drowsy and numbing quality. He recommended boiling the juice with hog's grease to make an unguent. This, he said, cured all inflammations and all manner of burnings or scaldings.

Hernández (1959, 3:67) reported that the Aztecs applied a decoction of the leaves of *D. stramonium* to the body for fever or administered it as a suppository. The fruit and leaves were considered good for pain in the chest. Instilled with water into the ears, *D. stramonium* was thought to alleviate deafness, and put into bedding it procured sleep for insomniacs. If too much was taken it was believed to cause insanity. It was thought to be 'cold' and lacked taste and odor.

In northwestern New Spain, the Ópata rubbed a leaf of *taguaro* on the painful area every morning before breakfast for "spleen disease." They believed it also matured tumors and abscesses (Nentuig [1764] 1977:62). Esteyneffer ([1719] 1978:871) said the seed was used as a purgative. Pfefferkorn ([1794–95] 1989:63–64) said the Indians misused it as an intoxicant disgracefully and that it killed some of them. There is evidence that the colonial Névome, and perhaps the aboriginal Névome, used *D. meteloides* to kill someone or to commit suicide (Pennington 1980:275).

Modern Use. Many species of *Datura* have been used in similar ways throughout the American and Mexican West. The Pima draw the pus from a boil by rubbing a green leaf of *D. discolor* with the thumb to make it slick, then applying the leaf to the boil. This also helps with hemorrhoids. Leaves are pounded with salt and placed on the sores. For earache, the flower bud is picked carefully so as not to lose the water or dew it contains; then it is heated and placed on an aching ear. If this treatment is applied too frequently, it is believed that there is danger of deafness. The buds are gathered early in the morning and kept fresh in a wet cloth; the liquid is then poured into sore eyes. Yaqui women were said to have formerly mitigated the pains of childbirth by drinking an infusion of the leaves. The Pima Bajo crush the leaves of a *toloache*, mix them with grease, and apply them as a poultice to bruises. The Mountain Pima toast the little fruit of *D. quercifolia*, mix them with grease, and apply

Datura

1. Used externally for pain and healing sores, boils, and hemorrhoids.

2. Effective pain killer; also has anticholinergic and antibiotic action.

3. Poisonous—contains hyoscine, atropine, hyoscyamine.

FIGURE 14   *Datura meteloides.*
a. Spiny seedpod. b. Seed.

them to sores. They place a leaf on a festering boil to make it burst and drain. The Tepehuan apply the leaves of *D. stramonium* as a poultice to the head for headache. For inflammations, they heat the leaves and then place them on the lesions.

The Mayo cook the entire *toloache* plant in water, then soak bandages in the mix and wrap them around the abdomen, or soak the legs in this brew for women in labor pain and difficulty. The women say that "cramps are cold and a woman must be warmed." To treat hemorrhoids, the green leaves are chopped and fried in vegetable oil until soft. A bandage is soaked in this mixture and is applied to hemorrhoids overnight. The Mayo also use *toloache* to attract or retain the attentions of a person of the opposite sex. The Warijio smear leaves of *D. wrightii* with animal fat or salve, applying this as a poultice to relieve aches, bruises, and sores. The Tarahumara apply leaves of a *toloache* to the forehead to alleviate headache but are careful to remove the leaves soon after the pain is relieved, or the person might "go crazy." They have many external medicinal uses for *uchiri*.

In Baja California Sur, hemorrhoids are soothed with an application of a decoction of the leaves of *D. discolor* that is also used for earache, rheumatism, and wounds. For pimples and to prevent inflammation, the skin is bathed with a decoction of the leaves or branches. The Paipai make a tea from the leaf of *D. meteloides* that they drink for venereal disease. The Seri make a tea from dry, mashed seeds of *D. discolor* to deaden the pain of a swollen throat.

The Hopi, Navajo, and Zuni all use *Datura* species. The Apache used the roots for chest trouble.

Today, Mexican Americans avoid the use of *toloache,* its reputation as a poison having spread. Periodically there are reported episodes of teenaged boys, experimenting with the effects, becoming temporarily blinded. Using jimsonweed as a hallucinogen is a major fad among teenagers across the United States today—a very dangerous practice because the dosage needed to produce hallucinations is very close to the lethal dose.

Phytochemistry. *Datura stramonium* contains hyoscine as well as atropine, hyoscyamine, apohyoscine, and meteloidine (Duke 1985:161–62). Thus it is poisonous and hallucinogenic but also an effective painkiller. In *D. metel,* scopolamine is the main alkaloid (Duke 1985:160). The scopolamine dilates the pupils of the eyes, blurring vision. The anticholinergic action dries the mouth and inhibits urination. Scopolamine and atropine have been synthesized for biomedical use. An extract of *D. discolor* branches was found to have a 2 + effectiveness against *Staphylococcus aureus* and *Streptococcus fecalis,* 1 + against *Bacillus subtilis* and *Candida albicans* (Encarnación and Keer 1991).

■ *Ephedra* (Ephedraceae)

MORMON TEA

Ephedra californica S. Wats.
chum wai, cañutillo, Paipai

Ephedra fasciculata A. Nels.
oo-oosti (sticks tea), *koopat,* Pima

Ephedra trifurca Torr.
oo-oosti, koovit nawnov, Pima; *popotillo, cañutillo del
     campo,* Spanish; Mormon tea, joint fir, English

Mormon tea is a small, essentially leafless herbaceous plant
with numbers of jointed needles, common in desert areas
of the American and Mexican West.

Ephedra

1. Used for urinary
complaints and venereal
disease.

2. Has diuretic, antispasmodic, and CNS stimulant
actions.

3. Because of pseudo-
ephedrine and tannin
compounds, it should be
used only under a physi-
cian's supervision.

Historic Use. *Ephedra* and *Equisetum* (q.v.), unrelated species, have been confused since antiquity. The Latin name
Ephedra translates as horsetail, derived from the Greek *epi-*
(upon) plus *hedra-* (seat), describing the plant's appearance;
the Spanish *cañutillo* emphasizes the hollowness of the
stem. *Ephedra* is known in Spanish as *cañutillo del campo,*
Equisetum as *cañutillo del llano.* Thus several peoples call
Ephedra species *cañutillo,* little tube, and give the same
name to *Equisetum* species, since both have hollow stems.
This is an ancient practice; Gerard in 1597 said that some
called a horsetail plant *ephedra.* In fact, another species of
Ephedra, used in Chinese traditional medicine for 5,000
years, is noted to have a slight resemblance to *Equisetum*
and in Japan as well as in China has been confounded with
the latter (Li Shih-chen 1973:161). Pliny (1938, 7:291) recommended it for cough, asthma, and colic.

Ephedra grows in arid, rocky lands and possibly has
been used since prehistoric times as a medicine in the American and Mexican West. The pollen appears in coprolites in
an ancient (A.D. 200–800) southwestern cave site (Reinhard, Hamilton, and Helvy 1991).

Modern Use. The Paipai use a tea from the leaves of *E. californica* for venereal disease. The Pima also take *Ephedra*
species in tea for venereal disease and treat sores, including
those caused by syphilis, by sprinkling powdered dry roots
on them. Indian tribes of Nevada used *Ephedra nevadensis*
and *Ephedra viridis* to treat venereal disease, for kidney dis-

orders, for delayed menstruation, and as a blood purifier (Train, Heinrichs, and Archer 1957:45). The Navajo use *Ephedra torreyana* S. Wats. for bladder and kidney problems as well as for venereal disease and afterbirth pains (Mayes and Lacey 1989:54), as do the Hopi and Zuni. Southern New Mexicans brew a handful of stems for kidney complaints. Anglos take Mormon tea as a beverage and also as a diuretic. Asthma sufferers who drink *Ephedra* claim that "it makes you feel spacy." It may be purchased in health food stores and supermarkets.

Phytochemistry. Like its related species in China, *Ephedra* has been used as an antispasmodic to dilate the bronchi in asthma even though, unlike the Chinese *Ephedra sinica*, it contains no ephedrine. Most authors state that *E. nevadensis* contains pseudoephedrine, also a potent stimulant of the central nervous system, which works on the heart and on the kidney. *Ephedra* is effectively used for urinary problems. It contains resins, much tannin, and a volatile oil (Duke 1985:177).

■ *Equisetum* (Equisetaceae)
HORSETAIL

Equisetum laevigatum A. Braun, also *Equisetum hyemale* L.
kawasíola, Tarahumara; *cañutillo del llano*, Colorado Spanish; *cola de caballo*, Spanish

Equisetum spp.
cola ce caballo, Tepehuan, Mountain Pima, Baja California Norte; horsetail, scouring rush, English

Equisetum laevigatum is a small, unbranched herbaceous plant with ridges along the stem and a spore-bearing cone at the top. It is found growing in moist alluvial or springy places. Both botanical (*equus*, horse; *setum*, tail) and common names reflect that the plant resembles a horse's tail.

Historic Use. This is one of the plants identified in a Neanderthal site dated at 60,000 years before the present. Galen said it had the faculties of binding and dryness without biting. Dioscorides reported many uses, including as a styptic

Equisetum

1. Used primarily for urinary problems, also for ulcers and as a vaginal douche.

2. Has antiinflammatory and antispasmodic action.

3. Poisonous to livestock, but there are no reports of human problems.

for nosebleeds and a drink for dysentery and to provoke urine, as well as for cough, penetrating wounds, and burns; these uses were also recorded by Pliny (1938, 7:375) and Gerard ([1633] 1975:1116). Its use was apparently not recorded for the Aztecs or Indians of northwestern New Spain.

Modern Use. The Tepehuan employ tea from the stems to relieve stomach cramps. The Mountain Pima make a drink from the flower stalk that they use for chest pain, and *Equisetum* is an ingredient of a lotion rubbed on inflammations. The Tarahumara use the wood to make a wash for cleaning wounds and to brew a tea for chest congestion. In Baja California Norte, several species of *Equisetum* are known as *cola de caballo* and are used for many purposes: problems of the kidneys, infections of the urinary tract, tuberculosis, arthritis, ulcers, and hemorrhoids.

Alaskan Yupik and Tanaina boil the upper parts of an *Equisetum* for pimples and sores (Fortuine 1988:193), as do Navajos, who also use the plants for various stomach problems (Mayes and Lacey 1989:52).

Cola de caballo is a popular remedy of Mexicans as well as Mexican Americans, who value its diuretic action. In Colorado, they make a tea from the stems for kidney ailments. A Tucson *curandero* recommends the tea for *mal de orín*, to clean the kidneys, and as a vaginal douche for women. It is sold in supermarkets as a medicinal.

Phytochemistry. Despite the many uses in folk medicine, phytochemists are dubious about its value. Most have considered *Equisetum* to be a weak diuretical though several varieties have been found as effective as hydrochlorothiazide (Pérez, Yescas, and Walkowski 1985). *Equisetum* is rich in silica and also contains the B vitamins nicotine, thiamin, riboflavin, and niacin, along with ascorbic acid (vitamin C), saponins, luteolin, and kaempferol. It is poisonous to livestock.

■ *Eryngium* (Apiaceae)
ERYNGO, BUTTON SNAKEROOT

Eryngium carlinae Delar.
saibari, Tarahumara

Eryngium hemsleyanum Wolff
yerba del sapo, Mountain Pima; *babadai, vasogadi,*
   Tepehuan

Eryngium heterophyllum Engelm.
yerba del sapo (toad herb), Mountain Pima; *sawíwarí,*
   Tarahumara

Eryngium longifolium Cav.
chinaca, Tarahumara

Eryngium maritimum L.
eryngium maritimum, sea holly, Renaissance English

Eryngium rosei Hemsl.
kukoida sivoradï, espíritu desilachado, Tepehuan

Eryngium spp.
segundo ocopiaztli, hoitzcolotli, espina de escorpión, Aztec;
   kokoida sivoradi (unraveled spirit), Tepehuan; *hutátci,*
   Tarahumara; *escorzonera, yerba del sapo* (toad herb),
   Spanish; eryngo, button snakeroot, English

Eryngium is a group of small herbaceous plants with spiny leaves arranged as a rosette around the base of the flowering stalk. Most grow in water or in marshy ground. In some species, the leaves have spines along the edges.

Historic Use. In the Old World, Gerard ([1633] 1975:1162) reported on a sea holly called *eryngium* in Latin and Greek: the roots were washed and boiled to treat kidney disease. Pliny (1938, 7:307) recommended it to counteract snake bites and all poisons, as well as for kidney problems, lumbago, dropsy, epilepsy, deficiency or excess in menstruation and all affections of the uterus, and various sores and eye problems (as an amulet). The Indian doctors of the Aztecs (Hernández 1959, 2:109) soaked the root of *Eryngium* for a brew that they gave to those who were convalescing, in order to "evacuate humors of fevers" through urine and sweat. *Eryngium* also helped those with joint illness, because it was believed to be 'hot' and 'dry' in the second degree. In northwestern New Spain, Esteyneffer ([1719] 1978:166) recommended using *Eryngium* in a compress for headache.

Modern Use. The Tepehuan drink a tea of the leaves of *E. rosei,* the Tarahumara a tea of the flower of *E. heterophyl-*

Eryngium

1. Eryngium has been used for heart palpitations, stomachache, and fever.

2. Some species have anti-inflammatory, estrogenic, and hypocholesteremic action.

3. Infamous abortifacient.

lum to relieve heart palpitations. From *E. carlinae*, the Tara-humara make a decoction by boiling the plant with salt; they drink the decoction for a cold and rub it on the head for headache. They use the flower and leaves of *E. hetero-phyllum* in a tea for eye treatments. They also use *E. longi-folium* for chest pains and cough. The Mountain Pima use *E. hemsleyanum* for bladder problems, and the Tepehuan use this species for stomachache. The Tepehuan use *E. het-erophyllum* in a drink for fever. They also eat the cooked flowers of *Eryngium humile* Cav. for colic and also the cooked stems and seeds of *Eryngium yuccaefolium* Michx. (common names unknown).

Escorzonera is an infamous abortifacient. I was told by a Mexican American woman that it was good to "bring down menstruation." It is sold in *yerberias* and grocery stores. Another plant with the name *escorzonera* is *Iostephane heterophylla*.

Phytochemistry. According to NAPRALERT, various *Eryn-gium* species contain active compounds. In a study using rabbits, *Eryngium campestre* was found to reduce the am-plitude of cardiac contractions and to decrease heartbeat. A study on rats using the aerial parts of *E. heterophyllum* showed hypocholesterolemic and estrogenic activity. The root contains coumarins, monoterpenes, and alkenynes. *E. maritimum* has antiinflammatory activity.

■ *Eucalyptus* (Myrtaceae)
EUCALYPTUS

Eucalyptus spp.
eucalipto, Baja California Norte, Spanish; eucalyptus, blue
    gum, English

Originating in Australia, these tall trees with long leaves and shredding bark are cultivated everywhere for their rapid growth and medicinal uses.

Modern Use. Introduced in the nineteenth century, *Euca-lyptus* is reported to be the eleventh most popular medici-nal in Mexico (Lozoya, Velázquez, and Flores 1988). The Mayo use it in tea for bronchitis. It is sold in all *yerberias* for use against colds, bronchitis, and other respiratory conditions.

Phytochemistry. The leaves contain tannin, primarily an essential oil composed of 70–80% cineol, as well as terpineol, sesquiterpene alcohols, and terpenes, with expectorant action on respiratory conditions (Font Quer 1979:398; Duke 1985:185). The gum contains tannins, catechins, and an antibiotic. The oil can cause fatalities if ingested in large doses, and to susceptible persons the leaves can cause dermatitis and respiratory allergic reactions.

Eucalyptus

1. Used for colds and other respiratory conditions.

2. Leaves contain essential oils with expectorant action; resin has antibiotic action.

3. In large doses the oil can cause fatalities and the leaves allergic reactions.

■ *Euphorbia* (Euphorbiaceae)

SPURGE

Euphorbia adenoptera Bertol.
golondrina, Warijio; *kueparim, golondrina*, Mayo

Euphorbia antisyphilitica Zucc.
candelilla, northwestern New Spain

Euphorbia cuphosperma Boiss.
picachali, Warijio

Euphorbia fendleri T. & G.
ch'il abe'é yáhi (little milkweed), Navajo; fendler spurge,
  English

Euphorbia indivisa Tidestrom
yerba de la golondrina, Mountain Pima

Euphorbia maculata L.
corape, Ópata; *golondrina, yerba de la golondrina*, Mayo,
  Spanish; spurge, swallow herb, English

Euphorbia melanadenia Torr.
natni xinak, golondrina, Paipai

Euphorbia misera Benth.
hamácj (fires), Seri

Euphorbia nutans Lag.
golondrina, Baja California Sur

Euphorbia plicata S. Wats.
kusí sigóname, Tarahumara; *jumete, candelilla china*, Yaqui

Euphorbia polycarpa Benth.

vee-ipkam, Pima; *golondrina,* Baja California Sur; *tomí-tom hant cocpétij* (land circular flat), Seri

Euphorbia trichocardia L. B. Smith
golondrina, Tepehuan

Euphorbia spp.
memeya, planta que mama leche, Aztec; *golondrina,* Tara-humara, Baja California Sur; *golondrina, yerba de la golondrina,* Spanish

Euphorbia species are remarkably diverse: some are small herbaceous plants, while others are large and superficially resemble cacti. All have milky white latex and small flowers. The genus *Euphorbia* is named for Euphorbus, physician to the king of Mauritania circa 85 to 46 B.C. The English common name spurge derives from *expurgare,* a Latin word meaning to cleanse, purify, cure. Many species are used in a multitude of cures everywhere, especially in Africa, where most of the five hundred species are endemic (Mitich 1983).

Euphorbia

1. Latex is used on warts and as an antivenom.

2. Depresses heart and respiration, destroys tissue, and is antibiotic.

3. Contains toxic compounds gallic acid, hydrocyanic acid, quercetin, and saponin.

Historic Use. In the Old World, Gerard ([1633] 1975:497–506) reported on spurges *Lathirys* and *Tithymalus* and the cactus *Euphorbium,* described by Hippocrates, Pliny, and Dioscorides. All were 'hot' and 'dry' in the fourth degree and caustic and were used in the manners described by the Aztecs and by peoples of the American and Mexican West.

Hernández (1959, 2:322–23) reported that there are many kinds of *memeyas,* the name Mexicans gave to plants that "secrete milk"; almost all of these were 'dry' and 'hot'. He criticized Mexican doctors, who believed them to be good for fever (which would contradict his 'hot' classification). He thought that because they "purge the pituita," *memeyas* could treat intermittent fever, consume growths of the eyes, calm toothache, cure both recent and old sores, close open skin, strengthen limbs, contain dysentery, and cure mange and wens. The latex was said to dissolve cataracts. This plethora of uses continues despite the toxicity of plants in this genus.

In northwestern New Spain, the Ópata prepared *golondrina* in a decoction or dried and pulverized it to bathe fresh wounds (Nentuig [1764] 1977:61). Esteyneffer ([1719] 1978:

838) said that the latex was good for hemorrhoids, to make an eye wash for cataracts, and to treat hydropsia. A strong decoction was good for the folk disease *tiricia* or *ictericia,* jaundice. Another *Euphorbia, candelilla,* was used for kidney stones.

Modern Use. The Pima use the juice of green *E. polycarpa* or make a tea for snakebite. They drink it, bathe in it, or apply it as a poultice to the wound, which causes vomiting and sweating. The Tepehuan boil the whole plant of *E. trichocardia* in an olla of water for ten to fifteen minutes, then cool it to apply to aching feet; they also use it for cleansing sores, wounds, and inflammations, they drink it as a tea for colic, or they poultice festering wounds with the sap. The Mountain Pima apply the sap of *E. indivisa* to an infection and boil the whole plant to make a lotion for wounds or ulcers.

The Tarahumara apply the latex of *E. plicata* to sores and wounds. They reportedly use the latex of another *Euphorbia* species for bruises, eye infections, inflammations, and wounds. The Mayo cook *E. maculata* or *E. adenoptera* in water, let the decoction cool, then bathe the skin for heat rash and boils. The decoction is also used to bathe the blisters of chicken pox. The cooked plant mixed with Vick's ointment may also be put on the blisters. The Warijio decoct the herbage of *E. adenoptera* to wash bites, stings, and sores. They report that they drop the raw, milky sap of *E. cuphosperma* into sore eyes—however, this is a dangerous action. The Yaqui use *E. plicata* as a drastic purge for diarrhea.

The Seri make a tea from the roots of *E. misera* to make a remedy for stomachache, dysentery, and venereal diseases. They mash the green leaves of *E. polycarpa,* adding salt and oil, then apply the mixture as a poultice to a swollen area. The same plant is also used in mixtures for toothache or heart pain. For snakebite, the Paipai wash the area with a decoction of *E. melanadenia* and if possible tie off the wound. In Baja California Sur, the bite of a snake or sting of a scorpion is washed with a decoction of the leaves or branches of a *Euphorbia* species; the decoction is also thought useful to wash the eye for conjunctivitis and the skin for infections. "Stomachache, urethritis, cystitis, kidney ache and colds" are all treated with the tea of a species in the Euphorbiaceae (Encarnación, Fort, and Luis 1987: 213).

Navajo mothers use the milky sap of fendler spurge to increase their milk supply, or for breast injuries; the Navajo also use the sap to treat indigestion, diarrhea, snakebite, and skin irritations such as poison ivy, warts, boils, and pimples (Mayes and Lacey 1989:120). The Hopi and Zuni make similar uses of *Euphorbia* species.

Mexicans and Mexican Americans say that they drink a tea of *golondrina* for diarrhea. They squeeze the fresh latex directly on warts. The leaves are boiled with a small piece of pomegranate, and this is administered as an enema twice daily to dry hemorrhoids. *Golondrina* may be purchased or pulled from gardens, where it is a weed.

Phytochemistry. *Euphorbia* species are poisonous. According to NAPRALERT, a study of *E. maculata* shows specific triterpenes and beta-sitosterol. *Euphorbia* species contain the toxic compounds dopamine, gallic acid, hydrocyanic acid, quercetin, and saponin (Duke 1985:555). These compounds may account for the destruction of warts. The latex is exceedingly irritating when applied externally. Common to many species of *Euphorbia* are diterpene (phorbol) esters, which are tumor promoting (Lewis and Elvin-Lewis 1977:121). *Euphorbia hirta* relaxes the bronchioles but also depresses respiration and the heart (Lewis and Elvin-Lewis 1977:299). Extract of the branches of *Euphorbia* cf. *polycarpa* Benth. have been found effective against *Staphylococcus aureus*, *Bacillus subtilis*, and *Candida albicans* (Encarnación and Keer 1991).

■ *Eysenhardtia* (Fabaceae)
KIDNEYWOOD

Eysenhardtia orthocarpa S. Wats.
koksigam, matariqui, Mountain Pima

Eysenhardtia polystachya (Ortega) Sarg.
coatli, tlapalezpatli, serpiente de agua, Aztec; *taray de Mexico,* northwestern New Spain; *palo dulce,* north-

western New Spain, Pima Bajo, Warijio; *si'palí*, Tarahumara; *báigno*, Mayo; *palo cuate*, Spanish; kidney-wood, English

Eysenhardtia species are medium-sized trees with fragrant foliage and white flowers. They grow on rocky hillsides in montane areas in northwestern Mexico.

Historic Use. Hernández (1959, 2:172) reported that the Aztec practice of drinking water in which some thorns of *coatli* had been soaked cleaned the kidneys, decreased the acidity of the urine, extinguished fever, and cured colic. He also said that the gum from this tree alleviated inflammations of the eyes and cleaned them of growths, which it could do because it was 'cold' and 'moist'.

In northwestern New Spain, Esteyneffer ([1719] 1978: 869) reported on a plant he called *taray de Mexico* (unrelated to the European *taray*, a *Tamarix*, which Dioscorides found to be paramount treatment for hydropsy [Font Quer 1979:287]). The water in which it was cooked was good for dropsy, fever, and liver and spleen disorders. It dissolved stone or sand in the kidneys or bladder. Tamayo in 1784 also said that drinking the water of *palo dulce* cured dropsy.

Modern Use. The Pima Bajo crush the bark of *E. polystachya* and steep it in water, which they apply as a lotion to the body to reduce high fever. The Warijio drink an infusion of the wood for stomach trouble. The Mayo take it for kidney problems. The Tarahumara make a tea from the bark for internal pain. The Mountain Pima mash and boil the roots of *E. orthocarpa* to make a drink for treating cough. Called *palo cuate* today, *E. polystachya* is sold in markets throughout Mexico and is used by Mexicans for treating kidney disorders. It may be purchased in at least one southwestern American supermarket chain.

Phytochemistry. Research conducted in Mexico, according to NAPRALERT, reveals only that *Eysenhardtia* is a weak hypoglycemic. Compounds include coatline A and B and other flavonoids in stem and trunk wood, beta-sitosterol in the bark and stem, and coumarin in the trunk bark.

Eysenhardtia

1. Used primarily for kidney disorders.

2. Has antiinflammatory, antitumor, and hypoglycemic compounds.

3. No reports of toxicity.

■ *Gnaphalium* (Asteraceae)

EVERLASTING, CUDWEED

Gnaphalium bourgovii A. Gray
manzanilla del río, Mountain Pima, Warijio

Gnaphalium canescens DC.
tzonpoton, tlacochichic, cauellos hediondos, Aztec (Ximénez)

Gnaphalium conoideum H.B.K.
chichictzompotónic, tzompotónic amargo, Aztec (Hernández)

Gnaphalium leucophyllum A. Gray, also *G. leptophyl-
lum* DC.
manzanillo del rio, Warijio

Gnaphalium macounii E. Greene
avo yoosigai (liviano flor), Tepehuan

Gnaphalium wrightii A. Gray
gordolobo, Mountain Pima

Gnaphalium spp.
cud-weed, cotton-weed, *gnaphalium*, Renaissance English;
manzanilla del rio, Tepehuan; *telempakate, rosábori*,
Tarahumara; *gordolobo*, Baja California Norte, Spanish; pearly everlasting, cudweed, English

Anaphalis margaritacea Benth. & Hook
gordolobo, Baja California Sur

Everlasting is a small herbaceous plant with woolly gray
foliage and flowers that have numerous tiny, pale yellow
to gray petals.

Historic Use. In the Old World, Pliny (1938, 7:443) said the
pale, soft leaves of *Gnaphalium*, given in a dry wine for
dysentery, arrested fluxes of the belly and excessive menstruation, was injected for tenesmus, and was applied to
festering ulcers.

Hernández (1959, 2:197) reported that the Aztecs used
the plant for curing ulcers. He said that the root should be
titrated for purging "phlegmatic humores safely by the superior route" (i.e., vomiting). Ximénez (1888:155), in the

version of Hernández's work that he published in 1615, said that a decoction of the *tzonpoton* plant "cured the chest" and that there was another kind called *tlacochichic* that had slender leaves, with the same uses. He stated that *tzonpoton* was 'hot' and 'dry' in the third degree.

Gerard ([1633] 1975:639–45) described twelve *Gnaphalium* species with the common name cudweed, the eighth of which had been named *Gnaphalium Americanum* by the botanist Clusius. He noted that the plant grows naturally near the Mediterranean Sea, but his botanical source Bauhin "affirmes that it grows frequently in Brasil" (p. 644). Gerard recommended that the smoke of the dried herb "prevaileth against the cough of the lungs, the great ache or paine of the head, and clenseth the brest and inward parts" (p. 644). He found these herbs to be of an astringent and drying quality.

Modern Use. The Tepehuan relieve heart pains and also coughing spells with a tea from the leaves of *G. macounii* and use another species to relieve stomach disorders or dysentery. The Warijio use two species of *manzanilla del río* as a stomach aid. The Mountain Pima take a tea made from the leaves of *G. wrightii* immediately before and after childbirth. The Tarahumara use the whole *Gnaphalium* plant in a tea for diarrhea and gastrointestinal problems as well as for cough. The Paipai use a plant, probably *G. wrightii* (no common name given), in a tea for cold and cough, a practice learned from non-Indians. In Baja California Sur, the species was identified as *Anaphalis margaritacea*. Here the flower, root, and branches are made into a decoction for a tea for cold and cough. It is part of a mixture of herbs that is taken as a tea before sleep. It is also good for bronchitis and fever. Mexican Americans in southern New Mexico and Arizona brew a tea, sweetened with honey, from the branches for a raspy voice. They take a tea of *gordolobo*, using sugar for sweetening, for a dry cough.

Much confusion has occurred over the common name *gordolobo*. In Europe, that common name was given to *Verbascum thapsus* (see chapter 2). *Verbascum* appears in the writings of Dioscorides, Pliny, and Galen as an herb that was useful for many conditions, especially chest conditions, menses, and hemorrhoids. *Verbascum thapsus* was natural-

Gnaphalium

1. Used for colds and coughs.

2. Has smooth muscle relaxant, hypoglycemic, and antiinflammatory action.

3. No reports of toxicity from correctly identified *Gnaphalium*.

ized to the cooler climates of the American and Mexican West, where it is called *punchon* and used to treat hemorrhoids. Because *Gnaphalium* has been widely confused with *Senecio* species, most of which contain pyrrolizidine alkaloids, the FDA banned *Gnaphalium* in the United States. However, in Mexico, *Gnaphalium* is one of two herbs that were distributed as a medicine by the Social Security Institute in 1984. Also, *Gnaphalium* was available in a Tucson herb store in 1994, as well as in Ciudad Juarez, Ciudad Chihuahua, and in Nogales, Hermosillo, Puerto Peñasco, and other cities in Sonora (see Kay 1994). As a final note, one wholesale company that sells widely in Tucson gives mullein, the common name for *Verbascum thapsus,* as the English common name for *gordolobo,* although the tiny ray flowers in the packet could not be *Verbascum* (see chapter 2).

Phytochemistry. Widely used, *Gnaphalium* species have been studied for medicinal use in many parts of the world. Some studies have shown activity as a hypoglycemic, cardiac stimulant, uterine stimulant, and smooth muscle relaxant, as well as antiviral, antiinflammatory, and antituberculosis activity. *Gnaphalium* and *Anaphalis* species have triterpenes, flavonoids, sterols, and tannins in the leaf and stem, as well as triterpenes and steroids in the flowers. The leaf of the Guatemalan species *Gnaphalium viscosum* has been shown to inhibit both *Streptococcus pneumoniae* and *Streptococcus pyogenes; Gnaphalium stramineum* inhibits *Staphylococcus aureus* (Cáceres et al. 1991).

■ *Guaiacum* (Zygophyllaceae)
LIGNUM VITAE

Guaiacum coulteri A. Gray
jullago, juyago, Mayo; *mocni, guayacán,* Seri

Guaiacum officinale L.
hoayacan, palo santo, matlalquáhuitl, Aztec; *guayacan, palo santo,* Spanish; lignum vitae, lignum sanctum, guaiacum, Indian pock-wood, English

Lignum vitae is a small tree with small resinous leaves and bright blue flowers, growing in tropical climates.

Historic Use. The epidemic of syphilis that followed Columbus's first voyage was early believed treatable with a medicine made from a tree called pock-wood or lignum vitae. Monardes (1574) was the author of the first description, which was quoted by Gerard ([1633] 1975:1611–12) from information given by the peripatetic Clusius: "It is of singular use in the French Poxes." It was judged to be 'hot' and 'dry' in the second degree, and a decoction of the bark was said to help cure "dropsie, asthma, epilepsy, diseases of the bladder and kidneys, pains of the joints, flatulences, crudities, and lastly all chronical diseases proceeding from cold and moist causes."

Hernández (1959, 2:394) noted that God gave this remedy for *mal gálico* (venereal disease) and that the tree also grew in Haiti. The extravagant claims were soon disproved. Nevertheless, the tree continued to be used as medicine, although rarely now. In northwestern New Spain, Esteyneffer ([1719] 1978:834) recommended it for dizziness, paralysis, the French disease, hiccough, hydropsy, melancholy, and many other disorders.

Modern Use. The Mayo cook the whole plant except for the root, putting the brew out overnight and using it daily for diabetes. The Seri crush the fruit of *guayacán,* then cook it in water to make a tea for dysentery. The resin makes a blue pigment that was sometimes used by the shaman.

In biomedicine, the resin is used as a reagent in tests for occult blood and was formerly used in the treatment of rheumatism (*Dorland's Medical Dictionary* 1974:670). It is used for various purposes in homeopathic medicine. Available in certain supermarkets, *palo santo* is made into tea for fever, and an aching head is covered with the leaves.

Phytochemistry. According to NAPRALERT, studies have found diuretic activity, activity against tuberculosis, and a menstruation induction effect. The stem bark contains daucosterol, guaiacol, and dehydroguairetic acid, as do the leaves and fruit. It was the diuretic effect that was once believed to cure syphilis. The leaf and stem contain alkaloids. The resin in the wood and fruit causes nausea and is poisonous if eaten in quantity (Lewis and Elvin-Lewis 1977: 279).

Guaiacum

1. Formerly used to treat venereal disease.

2. Causes diuresis and nausea, induces menstruation.

3. Resin is poisonous.

■ *Guazuma* (Sterculiaceae)

GUAZUMA

Guazuma ulmifolia Lam.

quauhólotl, árbol de élotl, Aztec; *ahiya*', Warijio; *axya*, Mayo; *huasima, guásima*, Pima Bajo, Tepehuan, Yaqui, Spanish

Guazuma is a shrub or tree up to twenty meters tall, with soft, serrated, whitish leaves hairy on the underside, and which has small, yellowish green flowers. The fruit is long and spiny. It usually grows in canyons and arroyos.

Guazuma

1. Tea is taken as a tonic.

2. Antibacterial, stimulant.

3. No reports of toxicity.

Historic Use. Hernández (1959, 3:76) said the Aztec *quauhólotl* was the same tree that the Haitians called *guazumo*. The fruit and bark of the trunk were 'cold' and glutinous, very efficacious for closing recent or old wounds. Oviedo, writing in the sixteenth century, said the Indians made a beverage of the fruit and that after a few days' use, they stopped getting fat (Martínez 1969:158).

Modern Use. The Pima Bajo use the bark in a tea for the kidneys, while the Tepehuan use this tea for shortness of breath. The Mayo cut the bark into small pieces, soak it in water for a day and a night, and use this instead of drinking water for diabetes. The Warijio eat the young fruit and use the mature seeds as a coffee substitute but apparently do not use any part of the tree as medicine. The Yaqui drink a tea made from the leaf for the blood. It is purchased for treating various conditions in *yerberias* and supermarkets in Arizona.

Phytochemistry. Probably because of the wide distribution of medicinal uses throughout Central and Latin America, various studies have been conducted there. However, NAPRALERT reports that little biological activity has been found except for cytotoxic and antibacterial activity against *Bacillus subtilis.* Caffeine was found in the leaf. Phytochemical screening of the bark showed absence of alkaloids, flavonoids, and saponins. Tannins were present.

FIGURE 15 *Gutierrezia sarothrae.* a. Central flower.

a

■ *Gutierrezia* (Asteraceae)

TURPENTINE BUSH

Gutierrezia sarothrae (Pursh.) Britt. & Rusby
collálle, New Mexico Spanish

Gutierrezia spp.
yerba de la vibora, Spanish; *escoba de la vibora,* New Mexico Spanish; snakeweed, turpentine bush, matchweed, English

Snakeweeds are small, resinous, often glutinous herbaceous plants with yellow flowers. They grow on dry stony plains, mesas, and slopes. Numerous species have been

Gutierrezia

1. Used for childbirth problems and for sore throat.

2. Has uterine stimulant and hemolytic activity.

3. Poisonous to animals, saponin content could cause human poisoning although none has been recorded.

given the common name *yerba de la víbora* or snakeweed: twelve overall in Mexico (Martínez 1979), including *Haplopappus heterophyllus* Gray (Pima), *Stellaria* species (Tepehuan), *Zornia* species (Mountain Pima), and possibly *Berlandiera* (Ópata).

Modern Use. Mexican Americans of New Mexico use *collálle* or *yerba de la vibora* for female complaints. In Santa Fe, the new mother would drink a few cupfuls of *collálle* to help involution of the uterus. In Las Cruces, they brew a tea from the branches, which they take four times daily until pain is gone: it "warms the ovary." The same tea is taken three times daily for sore throat with fever. The baby who has a cough is given two ounces of the tea with sugar three times daily. Drinking a tea or bathing in the decoction has been recommended for inflammation and pain, stomachache, and excessive menstruation (Moore 1979:74). *Gutierrezia* species are employed as medicines by the Navajo, Tewa, and Zuni. However, I could not find it listed as used by any other peoples of the American and Mexican West.

Phytochemistry. According to NAPRALERT, biological activity has been found in various species of *Gutierrezia*. Compounds isolated include many flavonoids, monoterpenes, diterpenes, saponins, and sesquiterpenes. Poisonous to animals, *Gutierrezia* species are potentially dangerous to humans because of the saponins.

■ *Haematoxylon* (Fabaceae)

LOGWOOD

Haematoxylon brasiletto Karst.
quamóchitl, hoitzquáhuitl, o brasil, Aztec; *sitagape,* Tarahumara; *huchachago,* Warijio; *brasil, palo del brasil,* Pima Bajo, Mayo, Yaqui, Baja California Sur, Spanish

Haematoxylon is a shrub or small tree with showy yellow flowers. Some individuals bear spines up to two centimeters long. The botanical name is formed from the Greek *haima* (blood) and *xulon* (wood), referring to the color of the wood. The common name in Spanish, like the name of the nation Brazil (named for the frequency of the tree there), reflects the red color of the wood, which has been used as a dye.

Historic Use. The Aztecs stopped diarrhea with the 'cold' and astringent plant the Spanish call *brasil* (Hernández 1959, 2:337), which has been identified by some as *H. brasiletto*. A decoction of *palo del brasil* treated *tirisia* (the name comes from icterus, or jaundice) in eighteenth-century northwestern New Spain. Esteyneffer ([1719] 1978:336) recommended giving "four or five live lice in a soft boiled egg" to the patient without his knowledge, repeating this for a few days. For severe and persistent jaundice, a decoction of the thorns was to be given.

Modern Use. The Tarahumara wash wounds with a decoction of the bark, which they also drink for diarrhea. In treating *tirisia* they use a bath of *brasil* water; then a few black lice are taken from someone's head, cooked in water, and drunk. The Pima Bajo boil the interior parts of a branch of *brasil* to make a lotion that helps small boys (never girls) when they are ill. The Yaqui use water in which a branch was soaked to cure rheumatism. The Warijio also use *brasil* medicinally. The Mayo treat depression with *brasil*.

In Baja California Sur, a decoction of the bark and branches taken as tea two to three times daily is believed to improve the circulation and to be good for heart trouble. Tea from the shoots with nutmeg and flowers of seven different citrus species is believed to control blood pressure and circulation.

Mexican Americans brew a tea of *brasil* leaves, which they both give to drink and put in bathwater for depression. Taking a bath in water colored red is believed to cheer the sad person. A new red dress gives the same results: it would appear that the effects of *brasil* are suggested by its magical red color. Brazilwood chips are sold as medicine in Arizona.

Phytochemistry. The heartwood of *Haematoxylon campechianum* L., related to *H. brasiletto*, contains 10% tannin (Morton 1981:321). An extract of the branches has been found to have 2+ activity against *Staphylococcus aureus* and *Streptococcus fecalis*. According to NAPRALERT, the

Haematoxylon

1. Used to treat depression and rheumatism, to wash wounds, and to improve circulation.

2. Has antibiotic and astringent action.

3. No reports of toxicity.

oxygen heterocycle brazilin shows activity against *Brucella abortus* and *Shigella flexneri.*

■ *Haplopappus* (Asteraceae)

JIMMYWEED

Haplopappus
[= *Aplopappus*]

Haplopappus heterophyllus (A. Gray) Blake
sai oos, Pima; *yerba de la víbora,* Spanish; jimmyweed, rayless goldenrod, English

Haplopappus laricifolius A. Gray
xal shaB u, Paipai; *hierba del pasmo,* Spanish

Haplopappus sonoriensis (A. Gray) Blake
hierba del pasmo, paroqui, Ópata; *romerillo amargo, hierba del pasmo,* Baja California

Haplopappus is a group of herbaceous plants and small shrubs with stiff, often resinous leaves and yellow flowers. Both *Baccharis* and *Haplopappus* species have been called *yerba del pasmo* and have been used to treat *pasmo,* infections that, according to folk belief, are caused by the body's rapid temperature change from hot to cold.

Haplopappus

1. Taken to treat the folk illness *pasmo;* also for cough, toothache, muscle pain.

2. Has antispasmodic, antiinflammatory, and antibiotic activity.

3. Contains the poisonous alcohol tremetol.

Historic Use. In northwestern New Spain, Ópata washed and ground a good handful of *yerba del pasmo,* put in an equal amount of boiling water, then beat and stirred the mixture constantly until thick to make a plaster on fractures. They fried the plant in oil or grease for an ointment to relax a spasm, for the 'heat' "instantly reduced the swelling of spasms." A decoction was given to newly delivered women who suffered from an internal *pasmo* (Nentuig [1764] 1977:61). The eighteenth-century Sonorans applied this or another *yerba del pasmo* to sores that had swollen or hardened so that they could not come to a head and burst, or if erysipelas had set in (Pfefferkorn [1794–95] 1989:64).

Modern Use. The Pima chew fresh leaves of *H. heterophyllus* to alleviate cough. To treat muscular pain, they warm a handful of leaves and apply to skin previously scarified with glass. Mexicans in Baja California Sur gargle a decoction made from the branches of *H. sonoriensis* for toothache. The Paipai make a tea of the leaves of *H. laricifolius* that they take for menstrual stoppage due to *frio,* cold. The Navajo and Hopi also use *Haplopappus* species as medicines for toothache, sore throat, and cough. Mexican Americans in Tucson make an infusion of the foliage of *H. heterophyllus* and drink it to treat *pasmo.* In Las Cruces, New Mexico, the infusion is taken for stimulating menses.

Phytochemistry. An extract of *H. sonoriensis* branches has been reported to show 2+ effectiveness against *Staphylococcus aureus* and *Bacillus subtilis* and 1+ against *Streptococcus fecalis* (Encarnación and Keer 1991). According to NAPRALERT, the leaf and stem of *H. laricifolius* contain apigenin, luteolin, quercetins, and kaempferol. *H. heterophyllus* contains tremetol, a toxic alcohol that has caused human poisoning (Fuller and McClintock 1986:317).

- ### *Heterotheca* (Asteraceae)
  TELEGRAPH PLANT, FALSE ARNICA

  Heterotheca grandiflora Nutt.
  miona, Mayo

  Heterotheca inuloides Cass.
  árnica, árnica del pais, Baja California

  Heterotheca subaxillaris Britt. & Rusby
  gordolobo, Warijio; telegraph plant, English

  Heterotheca spp.
  árnica, hárnica, Spanish; false arnica, telegraph plant, English

Arnicas are large, coarse, erect herbaceous plants with toothed leaves and yellow flowers. The common name comes from the Greek *arnakis* (lambskin), from the texture of the leaves.

Modern Use. The Mayo use the flower of *H. grandiflora* for cough. In Baja California Norte, a decoction of *árnica* is used to treat external and internal ulcers and lesions as well as poor circulation. The Mountain Pima make a lotion of

Heterotheca

1. Used externally to treat inflammations, wounds, and bruises.

2. Has antiinflammatory, antiseptic, and antimicrobial actions.

3. Contains toxic tremetol.

Heterotheca lamarckii (common name unknown) by boiling the plant for fifteen minutes, then applying the lotion to rheumatic joints. The Warijio use *H. subaxillaris* (which they call *gordolobo*) in various curative practices.

False arnica can be purchased in Mexican markets, for it is one of Mexico's most commonly used medicinals. The flower is used in a preparation to treat inflammations and bruises. Mexican Americans brew the leaves to make a wash for wounds. In Tucson it may be purchased at *yerberias* and grocery stores, where it is labeled *para uso external solamente* (for external use only). It is also available in American health food and herb stores.

Phytochemistry. Limited information is available on the chemistry of *Heterotheca* species (see Winkelman 1986:117). Plants in the genus contain sesquiterpene cadinene derivatives and mono-, sesqui-, and diterpenoids. In regular decoctions, taken internally, it is a highly toxic plant. It contains tremetol, which has caused stock losses and can make humans sick when they have drunk milk from cattle that have consumed *Heterotheca* (Huxtable 1983:186).

Because a common name for *Heterotheca* species is arnica, plants in the genus may be mistaken for *Arnica montana* L., a European species that is acclaimed when used topically for muscle spasms but is poisonous when taken internally, affecting the heart and circulatory system (Duke 1985:64). True arnica is used in homeopathic tinctures.

■ *Hintonia* (Rubiaceae)
COPALQUÍN

Hintonia latiflora (Sessé & Moc.) Bullock
[= *Coutarea latiflora* Sessé & Moc. = *Coutarea pterosperma* (S. Wats.) Standl.]
sif us, árbol amargo, Pima Bajo; civukali, Tepehuan; iwít-culi, Tarahumara; tapichowa, hutetiyo, Warijio; kupalkeen, Yaqui; copalquín, Mountain Pima, Mayo, Baja California Norte; copalquín, copalchi de jojutla, Spanish

Copalquín is a shrub or small slender tree with large white flowers. It does not grow in the United States and therefore has no common name in English.

Modern Use. The Pima Bajo make a tea of the bark, adding a bit of sugar since the bark is very bitter. This tea is drunk for fever and relief of influenza and is believed to enrich the blood. The Tepehuan make the bark into a tea to enrich the blood and also for fever, but in addition apply it as a lotion to cleanse wounds resulting from snakebite. The Mountain Pima soak small pieces of bark in cold water, which they afterward take as a fever remedy. The Mayo are reported to seek to improve circulation in diabetes by boiling the bark of *copalquín* with cinnamon and anise seed in water, adding sugar, and boiling the mixture to syrup. They take this syrup before each meal. Alternatively, they cook the bark in water, putting it out in breezes overnight to make a drink that is good for circulation of the blood. The Warijio take the bark tea as a purgative. The Tarahumara also use the bark in a tea for fever. A similar use is reported for the Yaqui, who also use it for diabetes. In Baja California Norte, *Hintonia* is recommended for diabetes, gall bladder problems, hepatitis, and malaria. The bark is widely used by Mexicans and Mexican Americans in the American and Mexican West, especially for fever and for diabetes. It can be bought at *yerberias*, grocery stores, and supermarkets.

Phytochemistry. *Hintonia* contains quinine, the common drug used for malaria, as well as alkaloids, flavones, glycosides, and saponins (Weiss 1988:277). According to NAPRALERT, *H. latiflora* contains coumarins in the stem bark, and animal studies show antimalarial and antihyperglycemic activity. *Hintonia latiflora* has been found to be among the most effective hypoglycemics (Pérez et al. 1984).

Hintonia

1. Widely used for diabetes and fever.
2. Has antimalarial and hypoglycemic activity.
3. No reports of toxicity.

■ *Ibervillea* (Cucurbitaceae)

COYOTE MELON

Ibervillea sonorae (S. Wats.) Greene
[= *Maximowiczia sonorae* S. Wats.]
guareki, Ópata; *wareki wareki*, Mayo; *hant yax*, Seri; *guareque, guareke, wereke*, Spanish; coyote melon, cowpie plant, English

Coyote melon is a small vine with a huge taproot. Shaped like a bottle, part of the root is aboveground. The plant, which may be found in various parts of Sonora and Sinaloa, generally dies back to the root during the dry season.

Ibervillea

1. Taken to treat cancer, cholesterol, and diabetes, and is used on wounds.

2. No studies of biological activity have been reported.

3. May have cytotoxic cucurbitacins.

Historic Use. Pfefferkorn ([1794–95] 1989:65–66) first described *guareke* as a tree "entirely bare and leafless": he was unaware that it was a vine. He praised its medicinal uses, stating that the root—dried, powdered, and sprinkled on—healed fresh wounds without causing pain. However, he said it was no good with a putrid or complicated wound. For the next few centuries, no other ethnography mentioned its medicinal uses.

Modern Use. The Mayo cut the tuber in pieces for juice to apply to rheumatic places. The tuber is used for arthritis, bruises, muscular pains, and wounds by the Mayo, Yaqui, and Seri as well as Ópata (López and Hinojosa 1988:44). Now Mexican herb stores feature *guareque,* the root sold in capsules to treat cancer, cholesterol, and diabetes, under which name it is also sold in Tucson. It is thought to "purify the blood." Fresh *guareke* is still desired for treating wounds, but it is difficult to obtain in the United States and is very expensive in Mexico City markets.

Phytochemistry. Early analyses indicate the presence of sterols, saponins, alkaloids, phenols, and other toxic compounds (Magdalena Ortega Nieblas and Ana Lilia Reina Guerrero, personal communication 1995). Since *guareke* is in the Cucurbitaceae, there is a potential for cytotoxic cucurbitacins to be present. This plant has not been studied sufficiently to warrant its extensive use.

■ *Jacquinia* (Theophrastaceae)
JACQUINIA

Jacquinia aurantiaca Ait.
segundo hoitzxóchitl, Aztec; *xuchipatli,* northwestern New
    Spain

Jacquinia pungens A. Gray
san juanico, palo san juan, Pima Bajo; *san juanico,* Wari-
    jio; *san juanico, tasiro takarn, jocojn, vichajowam tasiro,*
    Mayo; *cof,* Seri; *san juanico,* Spanish; jacquinia, English

San juanico is a small tree with thick, narrow, leathery leaves that are tipped with a long, stiff, yellowish spine.

Historic Use. According to Hernández (1959, 2:376), the Aztecs believed that they strengthened the heart by taking a decoction of the flowers. The pulverized bark could cure the eruptions of *mal gálico* (venereal disease). The brew also alleviated afflictions of the uterus. *Hoitzxóchitl* (*hoitz-*, spiny; *xochitl-*, flower) was said to be astringent, 'hot', and 'dry' in the second degree. In northwestern New Spain, Esteyneffer ([1719] 1978:309) recommended an enema of the decoction for colic, to counteract the 'cold' and 'moist' humor of flatus.

Modern Use. The Mayo in Las Animas, Sonora, mash the green fruit, which they apply to infected sores twice daily. For an earache, the fruit is roasted in ashes and chilled before its juice is expressed and strained. This brew is placed in the ear with cotton once daily. For a rash the fruit is cooked in water and drunk chilled. For a cold, the fruit is used in a tea. A tea of the bark is given for *alferecía* (convulsions) and paralysis. A tea of the flowers is given for asthma. For a cold, the Warijio make a paste from the seeds, form it into a ball, and insert it in the nose to cure catarrh. The Seri wash the face with an infusion of the flowers for dizziness and also use the infusion as eardrops. Pima Bajo use a tea of the flowers. Although *Jacquinia* is said to be an abortifacient, Mexican Americans do not report using it.

Phytochemistry. Little is known about the chemistry. Jacquinonic acid, an ant-repellent triterpenoid, has been isolated. The roots are also used to poison fish (Morton 1981: 653).

Jacquinia

1. Used for colds, asthma, and infected sores.

2. Has ant and fish poisoning activity from a triterpene.

3. Most likely toxic.

■ *Jatropha* (Euphorbiaceae)
LIMBERBUSH

Jatropha cardiophylla (Torr.) Muell.-Arg.
chichioaquáhuitl, árbol de nodriza, Aztec; *sangre de drago, sangre engrado, ensangregrado,* Spanish; limberbush, English

Jatropha cinerea (Ortega) Muell.-Arg.

sangre de drago, Pima Bajo; *lomboy blanco*, Baja California; *hamisj*, Seri; ashy limberbush, English

Jatropha cordata (Ortega) Muell.-Arg.

tzontecpatli, medicina de las heridas, Aztec; *wa'pe', torote papelio*, Warijio

Jatropha curcas L.

quauhayohuachtli, semilla de calabaza de árbol, Aztec; *matacora, mata muchachos*, Tarahumara; *sangre de drago, sangrengrado, sangre de cristo*, Spanish; physic nut, English

Jatropha dioica var. *sessiflora* (Hook) McVaugh
sangre de grado, Spanish

Jatropha malacophylla Standl.
[= *Jatropha platanifolia* Standl.]
ratowa, sangregrado, Tarahumara; *he'uho'*, Warijio

Jatropha vernicosa Brandegee
lomboy colorado, Baja California

Jatropha spp.
jicamilla de julimes, northwestern New Spain

Jatropha species are herbaceous plants, shrubs, or small trees with smooth limbs and sometimes with a rather acrid juice. The Spanish common names (usually versions of "dragon's blood" or "Christ's blood") come from the pale pinkish or brownish juice that exudes when the stem or bark is cut.

Jatropha

1. Used for weak blood, infections, wounds, and sores.

2. Drastic purge; shows activity against leukemia and other cancers.

3. The seeds are toxic, the sap irritating to the skin and toxic internally.

Historic Use. Hernández (1959, 2:55–56) said that the Aztecs used the plant now believed to be *J. curcas* as a purgative "of all humors, especially the thick and viscous, by all conduits but especially from above" (i.e., vomiting). Five or six nuts were toasted and dissolved in water or wine. He judged the plant to be 'hot' and greasy. Hernández (1960: 44) also said that the cut tops of "wound medicine" (*J. cordata*) had a latex that was pungent, bitter, 'hot', and 'dry' in the third degree, which when applied to recent wounds closed them to form scar tissue in a short time. Hernández (1959, 2:188) also wrote of a tree (*J. cardiophylla*) called "the

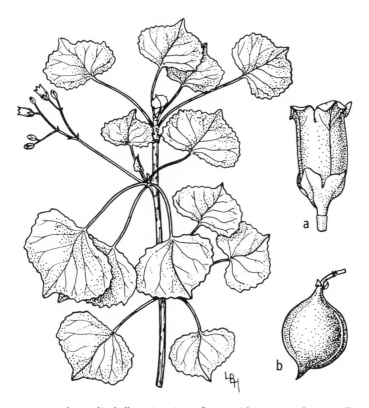

FIGURE 16 *Jatropha cardiophylla.* a. Staminate flower with a sympetalous corolla. b. Fruit.

tree of the wet nurse," so named because it secreted milk, that was said to lack flowers or fruit. Powdered bark of this tree, which he judged to be 'cold' and astringent, cured ulcers and produced cicatrization. In northwestern New Spain, *jicamilla de julimes,* tentatively identified as a species in the *Jatropha* genus, was used against the venom of snakes and other animals as well as arrow poison. This plant was also used to treat tear duct infections or fistulas, gum problems, phlegmons (inflammation of the cellular tissue), and fractures, and acted as a hemostat. Fray Juan Augustin Morfi ([1778] 1980:333), journeying to Texas, found himself in need of treatment for toothache, whereby he was recommended *sangre de drago.* It did not help, he thought perhaps because plants there were defective.

In territorial days of Arizona, plant remedies were

sought by the pharmaceutical house of Parke, Davis & Company. *Jatropha macrorhiza* was recognized as an excellent purgative and could be found in the 1890 catalog. It is a dangerous purgative, and the tubers have been eaten in error, possibly because of their resemblance to the edible jicama (Bender 1983:114–16).

Modern Use. The Warijio cut the stem of *J. cordata* and drop the juice directly into the eye to clear it and to treat eye diseases. The juice of *J. malacophylla* is applied directly from the cut twig end to cankers and other mouth sores. The Tarahumara use the latex of *J. curcas* for "bad eyes" and to relieve a toothache. They also apply the wood and latex of *J. malacophylla* to toothache. In Onovas, Sonora, the Pima Bajo boil *J. cinerea* leaves to make a lotion for aching limbs. Mexicans of Baja California Sur use a decoction of the bark or the sap of *J. vernicosa* to apply directly to hemorrhoids and to treat wounds. To treat an infection called *algodoncillo* ("little cotton," probably thrush), they wash the mouth with a soda solution followed by an application of the sap. The Seri use the roots to make a cure for dysentery: they remove the bark of the root from small plants, mash it, and make it into tea.

Mexican Americans make a decoction of the branches that they drink for weak blood. They may purchase the plant medicine in some Arizona supermarkets. A paper presented at a conference on traditional medicine in 1984 reported research that was conducted at the School of Dentistry in Zacatecas: successful treatment of dental problems (including loose teeth, gingivitis, bleeding gums, and toothache) with an extract of *J. dioica*.

Phytochemistry. Various uses for *J. curcas* have been reported, including extensive medicinal applications (Duke 1985:253–54). It is a strong purgative that can cause depression and collapse. The toxin is hydrocyanic acid. The leaves (which contain various beta-sitosterols, isovitexin, and vitexin) show antileukemic activity. Other species also contain antitumor compounds. The seeds contain the toxic protein curcin; nevertheless, the roasted seed is said to be nearly innocuous. The sap is irritating to the skin and toxic internally (Morton 1981:449). An extract of the branches of

J. vernicosa had 3 + activity against *Staphylococcus aureus* and *Bacillus subtilis,* and 2 + against *Streptococcus fecalis; J. cinerea* had 1 + against *Staphylococcus aureus* (Encarnación and Keer 1991). *Jatropha* is used by homeopaths in their microscopic doses.

■ *Juniperus* (Cupressaceae)
JUNIPER

Juniperus californica Carr.
choq, Paipai; *guata* (bark), Spanish

Juniperus deppeana Steud.
táscate, Mountain Pima; *awarí,* Tarahumara; juniper, English

Juniperus monosperma (Engelm.) Sarg. and *Juniperus communis* L.
sabino, northwestern New Spain; *sabina macha, sabina de la sierra,* Spanish; dwarf juniper, groundcedar, English

Juniperus osteosperma (Torr.) Little
ga'a, Mountain Pima; *gad bika'ígíí* (male juniper), Navajo; Utah juniper, desert juniper, English

Juniperus sabina L.
sabine, savin, English

Juniperus scopulorum Sarg.
gad ni'eetlii (drooping juniper), Navajo; Rocky Mountain juniper, English

Juniperus spp.
sabino, northwestern New Spain; *gayi,* Tepehuan

Junipers are common evergreen conifers that grow at higher elevations throughout Mexico and the southwestern United States. The bluish, berrylike cones are used to flavor gin. *Juniperus* was the name Virgil gave to the genus, from the Celtic word *juniperus,* rough (Mayes and Lacey 1989:55).

Historic Use. Gerard ([1633] 1975:1371–78), quoting Dioscorides, reported uses of various trees now grouped in the Cupressaceae family: juniper, cedar, sabine. All were 'hot'

Juniperus

1. Used for menstrual and urinary problems and body pain.

2. Increases intestinal and uterine contractions.

3. Highly toxic internally and externally; classic abortifacient.

and 'dry' in the third degree. The leaves of *sabino* when boiled in wine and drunk had many uses, including stimulating urination, for infirmities of the chest and coughs, and "to bring down the menses, draw away the after-birth, expel the dead child and kill the quick" (i.e., the living). The smoke of the juniper was good for cold and headache.

In northwestern New Spain, the branches or galls of *sabino* were made into a decoction. To break up kidney stones, an ounce or two of the drink was taken from time to time. However, this drink could not be given to pregnant women, because it was believed effective on the uterus. Women suffering from retained menses or childbirth problems were given an enema made with herbs that included *sabino* (Esteyneffer [1719] 1978:405).

Modern Use. The Mountain Pima make a drink from the branches of *J. deppeana* that they take for a cold. They burn *J. osteosperma* in the sickroom and inhale the smoke for headache. The Tepehuan make a sweat bath for ill persons using a *Juniperus* species. Branches are piled in the sleeping quarters of a house and fired; then water is thrown on the burning branches and a blanket positioned to drive the smoke toward the ill person, who is then covered with the blanket. To treat rheumatism, the Tarahumara make a wash from the needles of *J. deppeana* and also administer this in a vapor. They crush the needles to make a tea to treat body pain, sore throat, stomachache, and other problems. The Paipai steep the inner bark of *J. californica* until the fluid is red to treat pains in bones.

Mexican Americans in Colorado make a tea from the foliage of *J. monosperma* or *J. communis* that they drink when it is painful to urinate (*J. communis* is considered the more effective). For excessive menstruation, the branches are placed in a basin with hot water. The woman sits over it, covered with a blanket. Euro-Americans have used *Juniperus* as an abortifacient.

The Navajo use desert juniper as an emetic and in a medicine for headache, influenza, stomachache, nausea, acne, spider bites, and postpartum pain. They wrap a heated branch in a slightly damp cloth to relieve the aching of an ear, arm, or leg, or place it on the abdomen of a woman in labor (Mayes and Lacey 1989:55). They make a

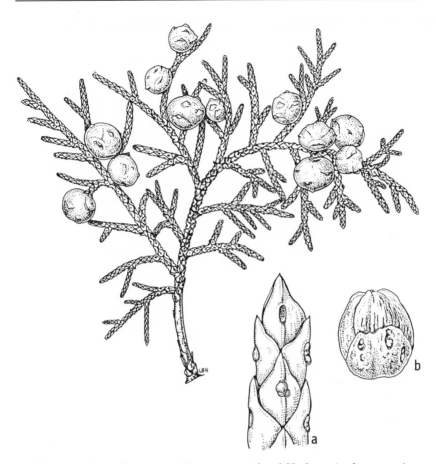

FIGURE 17 *Juniperus deppeana.* a. Mature appressed-scalelike leaves in alternate pairs, with glandular pits on the back. b. Seed.

tea of the needles of *J. scopulorum* to treat pain, stomach troubles, diarrhea, and spider bites (Mayes and Lacey 1989: 57). Teas made from the needles or berries of *Juniperus communis* are used by Alaskan Athabascans for chest pain, colds, sore throat, urinary retention, and tuberculosis. The Hopi and Tewa also use juniper as medicine.

Phytochemistry. A volatile oil composed of monoterpenes has been found in *J. communis* (Duke 1985:256). The diuretic principle is 4-terpineol, which is toxic. Externally the oil causes burning and inflammation of skin. It should not be used by expectant mothers because juniper and its extracts increase intestinal movements and uterine contractions. *Juniperus sabina* contains various essential oils. It is an energetic poison and causes abortion.

■ *Karwinskia* (Rhamnaceae)

COFFEEBERRY

Karwinskia humboldtiana (Roem. & Sch.) Zucc.
zazacatzin, zacate pequeño, Aztec; *cacachila,* Baja California, Tepehuan, Mayo; *himoli,* Warijio; *tullidora* (paralyzer), *cacachila, espinosilla,* Spanish; coffeeberry, English

Karwinskias are shrubs or small trees with round, bean-sized fruits that are mahogany red when mature and contain one or two seeds each.

Karwinskia

1. Used for fever and headache.

2. Effective antibiotic.

3. Has neuropathological toxins.

Historic Use. The Aztecs believed that the pulverized bark of the root of *zazacatzin* (*zacatl-,* herb; *tzin-,* small) made a gentle, harmless laxative and that it kept them well (Hernández 1959, 2:274).

Modern Use. The Tepehuan make a tea by boiling the bark of *cacachila* in a quart of water for several hours. This tea is taken hot or cold to reduce fever. The Warijio place the leaves on the forehead with salve to relieve headache. They find the fruits edible, but the seeds contain a poison. The Mayo mix the leaves with alcohol, then apply the mixture to the head for headache. The Mexicans in Baja California Sur make a decoction from the shoot or root that they take for fever and for headache. They know the fruit to be poisonous; eating it causes paralysis, which they treat by washing and massaging the paralyzed extremities with a decoction of the branches or roots of the same plant. Mexican Americans do not report its use.

Phytochemistry. According to NAPRALERT, neuropathological toxins have been found in the fruit and root of *Karwinskia* that caused flaccid paralysis of the hind limbs of animals, ascending to bulbar paralysis. Alkaloids were absent from the bark and leaves, and antimicrobial agents were present. Branches have been reported to have 3 + effectiveness against *Staphylococcus aureus* (Encarnación and Keer 1991).

■ *Kohleria* (Gesneriaceae)

TREE GLOXINIA

Kohleria deppeana (Schlecht. & Cham.) Fritch.
tlachichinole, tatachinole, Mayo, Spanish

Tree gloxinia is an herbaceous plant with woolly leaves and showy red to yellow-orange flowers.

Modern Use. This plant is not known by many ethnic groups, nor is it known pharmacologically, but it is used by many women in the American and Mexican West. The Mayo cook the twigs in water and leave this infusion in the open air overnight; they drink the infusion every morning for a week to treat urinary problems. Mexican and Mexican American women make a decoction of the leaves and flowers that they use in a vaginal douche for female problems such as leucorrhea and to relieve inflammation of the ovaries. They douche with an infusion of the leaves for menstrual problems, as recommended by a *curandero.*

Phytochemistry. According to NAPRALERT, no studies of *K. deppeana* have been reported, but the leaves, flowers, and sepals of the related Scottish species *Kohleria eriantha* contain the flavonoids apigenin, gesnerin, luteolin, and related glucosides and rutinosides. In 1952 an analysis of *K. deppeana* found 39.55% catechin tannins (see Martínez 1969:321–23).

Kohleria

1. Used for urinary and female problems.

2. High content of tannins suggests astringent action on mucous tissues.

3. No reports of toxicity.

■ *Krameria* (Krameriaceae)

RATANY

Krameria grayi Rose & Painter
cosagui, northwestern New Spain; *temitzo,* Ópata; *oeto,* Pima; *chacate,* Maricopa; *heepol,* Seri; *cósahui,* Spanish; white ratany, English

Krameria parvifolia Benth.
chacate, Tohono O'odham

Krameria spp.
wetahúpatci, Tarahumara; *cosawi,* Mayo; *cosahui,* Yaqui; *mesquitillo, crameria, clameria, grameria,* Spanish; krameria, ratany, English

Krameria is a small, gray, purple-flowered shrub believed to be parasitic on the roots of other desert shrubs. It is used for dye. The common name *chacate* comes from the Aztec *chacatl.*

Krameria

1. Used to strengthen the blood and for fever.

2. Some species have smooth muscle relaxant and antibiotic activity.

3. A Caribbean species is believed to be a major cause of esophageal cancer in that region.

Historic Use. It seems likely that the *temitzo* used in north-western New Spain when Nentuig ([1764] 1977:64) was in Sonora was a *Krameria* species. The Ópata dried and pulverized the root, applying it to fresh wounds. Longinos, traveling in Baja California in 1792, found that the root of a plant now tentatively identified as *Krameria* was used both as a lavage for diarrhea and to dye animal hides a beautiful scarlet. It was chewed for treating the teeth (Engstrand 1981:137).

Modern Use. The Maricopa chew the roots for sore throat. They make a strong and sour-tasting tea, taking one-half cup to reduce fever and for cough. After the newborn's cord is cut they pack the powdered root on the umbilical stump to give "absolute prevention of infection." *Krameria* is considered the most important medicinal plant for the Maricopa. Their neighbors the Pima boil the roots to make a decoction, which they drink or apply to sores, including those of "bad disease." The Tohono O'odham make an infusion of the twigs of *K. parvifolia* to treat sore eyes. The Mayo cook the leaves of a *Krameria* species in water, then cool and take the tea once daily for fever. The twigs are cooked to make a tea to treat ulcers and for circulation of the blood. The Yaqui cook the sprouts for a tea that they believe is good for the blood. The Tarahumara who live in the Chihuahua gorges pound the skin and mix it with suet. They apply this as an ointment to aching teeth. The Seri use the stems of *K. grayi* in a tea that will "make the blood very red." The flowers are prepared as a tea to cure the diarrhea of an upset stomach. The stems are dried and ground into a powder that helps healing and prevents infection of sores.

A Tucson worker who is more than eighty years old believes that the reason he can still climb up into the cathedral cupola to repaint it is that he takes *cosahui* every day. It can be found abundantly in a nearby park. *Krameria* also can be purchased in grocery and herb stores.

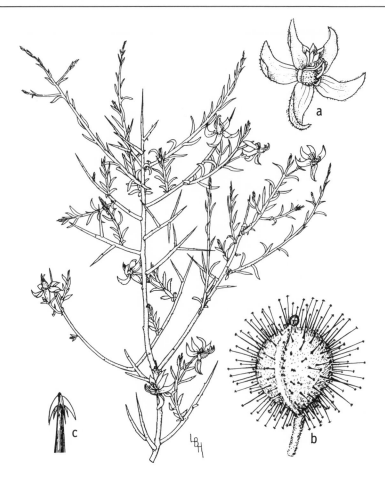

FIGURE 18 *Krameria grayi.* a. Flower. b. Fruit. c. Enlargement of the tip of a spine.

Phytochemistry. According to NAPRALERT, many analyses of *Krameria* species have been conducted. *Krameria cystisoides* exhibits antibacterial activity against *Staphylococcus aureus*. *Krameria tomentosa* has cardiac depressant activity as well as hypotensive activity and is a smooth muscle relaxant. *Krameria trianda* root shows antiviral activity; antibacterial activity against *Bacillus cereus, Bacillus subtilis,* and *Escherichia coli;* and antiyeast activity against *Saccharomyces cerevisiae. Krameria* species contain flavonoids and a high concentration of astringent tannins. A Caribbean species, *Krameria ixina,* is suspected of causing esophageal cancer (Morton 1981:356). *Krameria* species turn water red;

this may cue the folk medicinal uses for enriching the blood.

■ *Lantana* (Verbenaceae)

LANTANA

Lantana camara L.
[= *Lantana velutina* Mart. & Gal.]
piltzinteuhxóchitl (segundo), Aztec; *confitura blanca,* Warijio; *confituria,* Mayo, Spanish; lantana, English

Lantana glandulosissima Hayek
salvia real, Baja California

Lantana horrida H.B.K.
confiturilla negra, Pima Bajo; *confituria,* Warijio; *hamácj inoloj* (flames of fire), Seri

Lantana involucrata L.
pionia, uvalama, Tarahumara

Lantanas are small shrubs with showy flowers and black, drupe-like fruits. *Lantana camara,* which has yellow and orange flowers, is commonly cultivated as an ornamental. Several species of *Lantana* have been used as medicines throughout the tropical world, where they flourish.

Lantana

1. Used as an antivenom and for healing wounds.

2. Has antibiotic action.

3. Very poisonous, it is hepatotoxic.

Historic Use. *Lantana* is the Latin name for viburnum, which Gerard ([1633] 1975:1490) believed to be the same kind of plant in the Old World, for both were "low and bending shrubs." Gerard said the leaves and berries were 'cold' and 'dry' and of a binding quality. The decoction of either the leaves or the berries was good as a gargle for swellings, gum disease, and loose teeth.

The Aztecs made a vapor of the leaves of *L. camara* that they inhaled to remove the chills of fever and mitigate headache. Powdered it cured ringworm, rash, and leprosy if applied to the affected place or ingested. Hernández (1959, 3:87) judged it 'hot' and 'dry' in the third degree.

Modern Use. The Pima Bajo boil the leaves of *L. horrida* for a short time to make a lotion for healing wounds. The Warijio use a decoction of *L. horrida* or *L. velutina* to wash insect stings and snakebites. The Tarahumara pound and cook the roots of *L. involucrata* to make a drink that cures an "impacted stomach" or acts as an aid in childbirth. The Mayo boil the leaves of *L. camara* in water, applying the cooked leaves to snakebites. The Seri put the fruit and leaves of *L. horrida* in boiling water, then wash the head with this infusion to treat dizziness. The roasted leaves of *L. glandulosissima* are applied by Mexicans of Baja California Sur to painful areas to assist in cicatrizing. The decoction is also a remedy for cough, rheumatism, earache, and headache; a decoction of the branches of *L. camara* is used for cold and cough. Mexican Americans do not report using *Lantana*.

Phytochemistry. *Lantana* species have been widely studied for medicinal use, especially *L. camara*, as indicated by NAPRALERT reports. (Other species used in the American and Mexican West do not have this wide distribution of analyses.) When taken orally, *L. camara* is toxic: it has been shown to have hepatotoxic activity, causing obstructive jaundice, photosensitization, and a rise in serum glutamic-oxaloacetic transaminase as well as smooth muscle stimulation in animals.

In a study of *L. camara* in Baja California, an ethanol extract of branches was reported to have 3 + activity against *Staphylococcus aureus* and *Bacillus subtilis* and 4 + against *Streptococcus fecalis*, but *L. glandulosissima* showed no antibacterial activity (Encarnación and Keer 1991). According to NAPRALERT, studies in other countries are conflicting, most showing *Lantana* species to have antibacterial activity against *Staphylococcus aureus* and *Pseudomonas aeruginosa* but not against *Bacillus cereus* and *Escherichia coli*, and no antifungal activity.

The aerial parts of *L. camara* contain numerous triterpenes in addition to caryophyllene, phellandrene, lantatine, linalool, cineol, eugenol, and beta-sitosterol; the root bark contains tannins. Human poisoning has been attributed to ingestion of green fruits (Duke 1985:267).

■ *Larrea* (Zygophyllaceae)

CREOSOTE BUSH, GREASEWOOD

Larrea tridentata (DC.) Coville
[= *Larrea divaricata* Cav.]
cubiasisi, Ópata; *shoegoi*, Pima; *kovanow*, Yaqui; *uB shiih*,
    Paipai; *haaxat*, Seri; *gobernadora*, Baja California
    Norte, Spanish; *hediondilla* (Sonora, Pima Bajo), *gua-
    mis* (Chihuahua and New Mexico), Spanish; creosote
    bush, greasewood, chaparral, English

Creosote bush is one of the most common shrubs in the
flatlands of the Sonoran Desert. It has shiny, resinous, two-
lobed leaves; small yellow flowers; and fuzzy, white fruits.

Larrea

1. Used for headache, for
kidney and high blood
pressure problems, and for
cancer, sores, and pain; a
virtual panacea.

2. Has antimicrobial and
analgesic action.

3. May be linked to
hepatitis.

Historic Use. This New World bush is a veritable panacea
for many in the American and Mexican West, particularly
those who live in very arid regions. Pollen from the plant
was found abundantly in coprolites of Archaic southwestern sites (Reinhard, Hamilton, and Helvy 1991). *Larrea* seems to have been used throughout this area but not
outside of it: no plant described by Hernández has been
identified as *Larrea*, nor does it grow in the areas that he
surveyed.

A few eighteenth-century writers in northwestern New
Spain described its medicinal use. Pfefferkorn ([1794–95]
1989:65), mainly scornful, noted that "the doctors in
Sonora (that is, the old Spanish women) make use of this
herb in treating different illnesses" but concedes that it was
especially powerful for worms in children. Nentuig ([1764]
1977:63) noted that an ointment made from *cubiasisi* was
efficacious for gnarled, rheumatic limbs. Longinos in the
late eighteenth century found that the plant in Baja California was used to induce abortions and facilitate delayed
menstruation and expulsion of the afterbirth (Engstrand
1981:137).

Modern Use. The Pima Bajo grind the leaves of *hediondilla*
and apply them to aching teeth or extract the juice by crush-
ing the leaves between two fingers and rubbing them on the
gums. The Pima chew and swallow the gum for dysentery.
A handful of the ends of green branches are added to a pint
of cold water, which is then boiled for twenty minutes,

strained, and cooled for a drink that relieves gas or a headache caused by an upset stomach. The heated branches are
applied to painful areas. A decoction is drunk for tuberculosis. Fever is treated by using the warm drink as an emetic.
An infusion of the leaves is applied as a lotion for the sores
of impetigo. The Maricopa massage the infusion into the
scalp for dandruff and sprinkle the powdered dry leaves
into the armpits as a deodorant or on the feet to prevent
them from perspiring (Curtin 1949). The close cousins of
the Pima, the Tohono O'odham, use *L. tridentata* in the
same ways. The Yaqui use *Larrea* for arthritis, stomach ailments, and kidney problems, and as a deodorizer.

The Seri make much use of *haaxat.* They make a drink
by placing a wad of the lac (an encrustation on the leaves
deposited by insects), on a stick and then heating it in a
flame. The drops of lac that melt off the stick are caught in
a container of water, which is then drunk for contraception. The heated branches are used for various pains. For
headache, the Seri wrap hot leafy branches in a cloth and
hold them to the head or wash the head in water in which
the foliage was cooked. Aching feet are treated by placing
them on a cloth that was put on branches heated on hot
coals. A sore leg is held into smoke coming from a pit of
heating branches. The woman in childbirth is instructed to
lie over a small pit on which leafy branches are heated. The
heated branches serve as a poultice to alleviate abdominal
pain after delivery of the placenta. For breathing difficulty,
the patient is made to vomit in the morning on an empty
stomach, then is given a tea made of the leafy branches.

All these uses for this plant and more occur in both Baja
Californias, including for purifying the blood as well as
for kidney problems, gallstones, urinary tract problems,
rheumatism, arthritis, diabetes, wounds, skin problems,
tumors, uterine displacement, frigidity, and paralysis. The
root, branches, or bark are decocted for an abortive but
also to promote conception. The Paipai of Santa Catarina
boil the seedpods in water to make a wash for "painful
members." They drink this tea for night cough and also use
it for stomach trouble.

Mexican Americans in New Mexico steep the branches
to make a tea taken for kidney problems, *pasmo,* and high
blood pressure. They treat joint pain with either a brew of

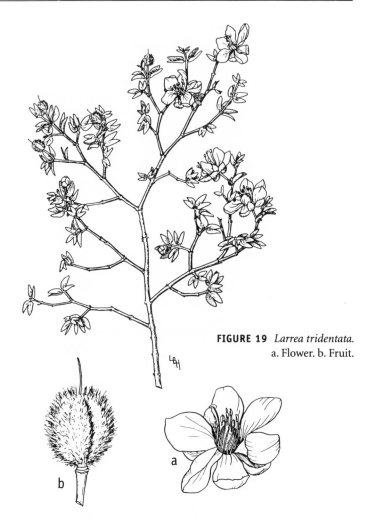

FIGURE 19 *Larrea tridentata.*
a. Flower. b. Fruit.

the leaves or direct application of the leaves, which are held in place with a bandage until they are dry. The leaves are placed on a painful pimple, a *postemilla*. For skin cancer, the leaves are rinsed and dried, pulverized, and brewed into an infusion that is used to wash the sore before it is bandaged. Mexican Americans drink a tea made from only a few of the bitter-tasting leaves daily before breakfast for uterine cancer. It is available in some supermarkets but most likely is collected outdoors.

Creosote bush has also been used by Euro-Americans. It has been touted as a cancer cure. As a child, I was given it in a steam tent for bronchitis.

Phytochemistry. Nordihydroguaiaretic acid (NDGA), which is the main phenolic constituent and a potent antioxidant, is present, as are more than a hundred other natural products such as flavonoids, volatile oils, wax, esters, and triterpenes (Timmerman 1981). The phenolic resin constituents, including NDGA and flavonoids, have been shown to be biocidals with antiseptic, antimicrobial, and bactericidal action. In addition, NDGA has analgesic properties and antitumor activity. In antimicrobial screening, an extract of branches was found to have 3+ activity against *Staphylococcus aureus*, 3+ against *Bacillus subtilis*, and 2+ against *Streptococcus fecalis* (Encarnación and Keer 1991).

Recently, five cases of hepatitis in the United States have been associated with use of chaparral tea. This led both the American Herbal Products Association and the FDA to ask that it be removed from the shelves of health food stores, although extensive chemical and pharmacological studies have failed to find any hepatotoxic properties in the tea (Blumenthal 1993b:93–94). In February 1995, the AHPA rescinded the chaparral ban, creating a hotline ([512] 469-6344) for reporting adverse effects (HerbalGram 1995, vol. 35:23).

■ *Ligusticum* (Apiaceae)

LOVAGE

Ligusticum porteri Coult. & Rose
  payihosa, Mountain Pima; *guariaca, chuchupate, chuchupaste,* Tarahumara; *osha,* Pueblo; *osha de la sierra, chuchupate,* Spanish; lovage, English

Ligusticum spp.
  acocotli tepecuacuilcense, Aztec; *Levisticum vulgare,* common lovage, Renaissance English; *chuchupate,* northwestern New Spain; *la pudosa,* Ópata

Lovage is a fragrant plant with highly divided leaves and white flowers that grows in the shade of pine-oak forests. *Levisticum* and *Ligusticum,* which are now names assigned to related genera, were both given for the same plant in the seventeenth century (Gerard [1633] 1975:1008). The Latin name was derived from the Italian province of Liguria; the

Spanish name *chuchupate* is modified from the Nahuatl *chichipatli*. The name *oshá* is applied by the Pueblo Indians to *L. porteri.*

Ligusticum

1. Used for fever, colds, and infection.

2. Probably has smooth muscle relaxant, antiin-flammatory, and fungicidal activity.

3. No reports of toxicity.

Historic Use. In the Old World, Pliny (1938, 6:99) said *Ligusticum* was good for the stomach, convulsions, and flatulence; Dioscorides stated that it moved urine and menstrua. Gerard ([1633] 1975:1008) said *Ligusticum* was 'hot' and 'dry' in the third degree, with the root used for all inward diseases, the distilled water as a wash for skin problems.

The Aztecs (Hernández 1959, 2:9) sprinkled the powdered leaves of one *Ligusticum* species on chronic sores to cure them, applied powdered leaves as a plaster to swollen legs, and washed the mouth ulcers with its juice; they used the root of another *acocotli* to treat stomach complaints by eliminating gas and mitigating stomachache, evacuating urine, provoking sweat, and driving out 'cold'. Hernández judged this plant 'hot' and 'dry' in the third degree. Several *acocotli* are identified as likely *Ligusticum* species (Valdés and Flores 1984).

In northwestern New Spain, it was employed at Rio Chico in colonial Sonora for stomachache (Pablos 1784). Nuñez (1777) wrote in the annual report to his provincial that for pain proceeding from 'cold', the root was boiled and the resultant tea drunk, while the remainder of the root was chewed and applied with the saliva to the painful part.

Modern Use. The Tarahumara use a tea of the roots of *chuchupate* for flatulence and gastrointestinal upsets. It is applied as a plaster to alleviate rheumatic pains. The Mountain Pima make a tea from the root for colds and fever.

Mexican Americans in Colorado bathe a person with fever in an infusion of the root. The root, dried and ground into a powder and applied in a gauze to wounds, will aid in healing by preventing infection. Northern New Mexicans use the root in many medicines. In Arizona the Mexican Americans sniff the root to clear a stuffy nose (it has a pleasant, spicy fragrance) caused by a cold. Children would dread telling their mother that they had *hinchazón* (distention of the belly), *bilis* (bile, meaning irritable), or constipation because she would treat any one of these conditions with the bitter tea.

Phytochemistry. According to NAPRALERT, no studies of biological activity or compounds of *L. porteri* have been reported. Related species contain coumarin, as well as flavonoids, acetylenic compounds, terpenoids, and essential oils (Bye 1986:115). Phthalides have been isolated from *Ligusticum* species; these may inhibit smooth muscle activity and account for its use in gastrointestinal complaints. *Ligusticum* is reported to contain compounds that may fight ticks and fungi (Begley 1992).

■ *Lippia* (Verbenaceae)
OREGANO

Lippia dulcis Trevir.
cococxíhuitl ocimino, Aztec; *yerba dulce,* Spanish

Lippia formosa Brandegee
oregano, Baja California Sur

Lippia origanoides H.B.K.
oregano, Spanish

Lippia palmeri S. Wats.
oregano, Baja California Sur; *xomcahiift,* Seri

Lippia pringlei Briq.
[= *Lippia umbellata* Cav.]
coapatli origanoide, Aztec; *bacatón, talakao,* Warijio; *bacot tami, colmillo de vivora* (snake's fang), Mayo

Lippia spp.
oregano, Spanish

Lippia species are small shrubs, many of which bear the common scent of oregano. The Spanish common name *oregano* is applied to nineteen different species representing ten genera in Mexico—this presents a serious and (unfortunately not unusual) problem that should make the user very cautious. In the American and Mexican West, the medicinal plant may be a Lamiaceae, such as *Hedioma* or *Monarda,* but most likely it is a *Lippia.* The culinary herb that has the common name *oregano* in English is *Oreganum vulgare* L.

Lippia

1. Used for colds and delayed menstrual periods.

2. Has antibiotic activity.

3. No reports of toxicity.

Historic Use. The various *Lippia* species have a long history of medicinal use in the Old and New Worlds. Gerard ([1633] 1975:667) reported that "all the organies" are remedies against the bitings and stings of venomous beasts. The decoction "provoketh urine, bringeth down the monthly courses, looseth the belly, is against old cough."

According to Hernández (1959, 2:208), the Aztecs made a mash of the root and leaves of *cococxíhuitl ocimino* (*coco-*, pain; *xihuitl-*, herb; *ocimino*, leaves resembling *Ocimum* [q.v.]), a plant now identified as *L. dulcis*. This mash was injected into the rectum or taken with a small amount of water to cure colic, flatulence, stomach and belly pains, and all 'cold' afflictions. It was 'hot' and 'dry' in the fourth degree. Hernández (1960:77) said that an *oregano* (now believed to be *L. pringlei*), 'hot' and 'dry' in the third degree, relieved flatulence when applied in any manner.

In northwestern New Spain, Esteyneffer ([1719] 1978:163, 855) recommended a decoction of *oregano* given as an enema or a drink for headache, counteracting the causative superfluity of the humor "pituita." He also used it for other conditions including vomiting, lack of appetite, paralysis, deafness, and stomachache.

Modern Use. The Warijio use the sap of *bacatón* to treat a toothache. To soothe a headache or treat a bruise, the leaves are steeped in hot water, coated with mentholatum, and applied to the head or bruise. The Mayo wash the sores of insect stings and snakebites with *bacot tami* cooked in water and take it twice daily for cough. Baja Californians Sur make a decoction of the flowers of *L. formosa*, drinking the tea for a cold and also bronchitis. For treating cold and cough, they make a decoction from the branches and leaves of *L. palmeri*. This decoction is also said to be good for the colic of a newborn infant. The Seri also use *L. palmeri*, making an infusion by putting fruit and leaves in boiling water. To treat dizziness, they wash the head with the infusion.

The Mexican child is helped to sleep when fresh leaves of oregano are placed under his pillow. Mexican Americans in Tucson are most likely to use culinary oregano, *Oreganum vulgare* L., rather than the *Lippia* species described above. They drink an infusion of the leaves of oregano for a cold. They also use it as an emmenagogue when the menstrual period is "late." Culinary oregano is sold in herb stores as well as grocery stores.

Phytochemistry. According to NAPRALERT, studies have not included those species of *Lippia* used in the American and Mexican West except for *L. dulcis* in Mexico, which contains ascorbic acid, monoterpenes, sesquiterpenes, and a newly identified sweetener named hernandulcin. Extract of *L. formosa* branches is reported to have 4+ activity against *Staphylococcus aureus,* 3+ against *Bacillus subtilis,* and 2+ against *Streptococcus fecalis;* extract of *L. palmeri* branches is 2+ effective against *Staphylococcus aureus,* 1+ against *Bacillus subtilis,* 2+ against *Streptococcus fecalis,* 2+ against *Escherichia coli,* and 2+ against *Candida albicans* (Encarnación and Keer 1991). Leaf of the Guatemalan species *Lippia alba* inhibited growth of *Staphylococcus aureus, Streptococcus pneumoniae,* and *Streptococcus pyogenes; L. dulcis* inhibited growth of *Streptococcus pneumoniae* (Cáceres et al. 1991). In vitro screening of *L. dulcis* demonstrated activity against *Salmonella typhi* and *Shigella flexneri* (Cáceres et al. 1993).

■ *Lysiloma* (Fabaceae)
FEATHERBUSH

Lysiloma acapulcensis Benth.
tepehoaxin, hoaxin del monte, *tepemizquitl,* Aztec; *matze, tepeguaje,* northwestern New Spain, Ópata

Lysiloma candida Brand.
gokio, kokio, palo blanco, Cochimíes

Lysiloma divaricata (Jacq.) Macbr.
wapakuwe, Tarahumara; *mauuta, sahi,* Warijio

Lysiloma schiedeana Benth.
mauto, ma'af, Pima Bajo; *wapakuwe,* Tarahumara

Lysiloma watsoni Rose
masav, tepeguaje, Pima Bajo; *guido sakoi* (large *tepeguaje*), Tepehuan; *mechowí,* Tarahumara; *machawi,* Warijio; *tepeguaje,* Mayo

Lysiloma spp.

tepeguaje, Spanish; featherbush, English

Featherbushes are large, attractive, thickly foliated shrubs with white flowers. The Nahuatl name comes from *tepetl,* hill, and *hoaxin,* plant that produces pods. Plants with *hoaxin* incorporated in the Nahuatl name all have pods and are classified in contemporary science in the family Fabaceae.

Lysiloma

1. Used to freshen the stomach and to fix loose teeth.

2. Bark has tannins with astringent and antibiotic activity; seed has hemagglutin activity.

3. No reports of toxicity.

Historic Use. Hernández (1959, 2:131) wrote that the Aztecs treated fever with a purgative made with the bark of *tepehoaxin.* He judged it bitter, mucilaginous, and 'hot'.

In northwestern New Spain, Esteyneffer ([1719] 1978: 239, 248, 506) recommended cooking the bark of *tepeguaje* in water, adding a little vinegar and honey, and using this as a gargle. Because it was astringent, it would fix loose teeth as well as treat "fallen uvula" and protect against smallpox. The Ópata cooked the inside of *Lysiloma* bark in water to make a wash that was said to clean dirty sores by maturing and filling them. In the settlement of Rio Chico, Pablos (1784) noted that sipping water in which the bark had been cooked was good for sores in the throat. Tamayo (1784) in Arivechi said the bark made a purgative for pain in the bones. The Cochimíes of Baja California treated sores with *gokio:* they dried and ground the bark into powder, then applied it to the wound (Barco [1768] 1973:66).

Modern Use. The Pima Bajo mash the bark of *mauto,* mix it with sugar, and apply it to the lids of sore eyes as treatment. They also mash a bit of the bark to make a poultice that they hold on an aching tooth with the fingers to relieve the pain; they also steep the bark in cold water to make a decoction for cleaning the teeth. The Tepehuan hold the boiled, cooled scrapings of the bark of *guido sakoi* in the mouth for sore gums. The Mayo chew the bark of *tepeguaje* to fasten loose teeth. The Warijio chew the bark of *machawi* for ailing teeth and gums and to tighten the teeth. They make a strong, bitter decoction for a mouth rinse by boiling the bark in water. The Tarahumara have the same practice, and they also use the bark in a tea for venereal disease. *Lysiloma divaricata* is similarly used. A species of *Lysiloma* is sold in Mexican herb stores, recommended for "refreshing or cooling the stomach," and it can also be purchased in some Arizona drugstores and supermarkets.

Phytochemistry. According to NAPRALERT, *L. acapulcensis* bark contains tannins. Phytochemical screening found that alkaloids, flavonoids, and saponins were absent. Some antibiotic activity was also demonstrated. The seed extract of *L. watsoni* has hemagglutin activity, inactivated by heat. If this change due to heating is true also of the bark (although no such test has been reported), chewed bark may be more effective than boiled.

■ *Malva* (Malvaceae)
MALLOW

Malva parviflora L.
   tashmahak, Pima; *malva,* Tarahumara; cheese weed, English

Malva rotundifolia L.
   [= *Malva neglecta* Wallr.]
   malva, Tepehuan, Mountain Pima, Spanish

Malva spp.
   alahuaccioapatli, aalacton, medicina mucilaginosa, hierba mucilaginosa, Aztec; *malbas simarronas* (wild mallow), northwestern New Spain; *malva, malba,* Spanish; mallow, English

Mallows are commonly found erect to trailing weeds, with roundish leaves and small flowers.

Historic Use. Mallow has been known in the Old World since the eighth century B.C. as a widely used medicinal. Dioscorides said that besides relaxing the belly, mallow is useful for the intestines and the gall bladder (Font Quer 1979:405). The leaves help against bee and wasp stings. Pliny (1938, 6:129–35) is lyrical in praise, noting its help with childbirth and treating sores.

   Hernández (1959, 3:223) wrote that the Aztecs recommended mashing the stems of a mallow and taking this with water, to help with a difficult delivery and mitigate the pain. Because mallow was 'cold' and mucilaginous, it was used to calm the pain of the "French disease," syphilis.

Malva

1. Used to wash out the kidneys, for stomachache, and to treat wounds.

2. Mucilaginous and mildly astringent. Leaves are edible and high in vitamin C.

3. No reports of toxicity.

Hernández (1959, 2:148) described another *Malva* "similar to *Malvavisco* in form and nature" that combated the acidity of the urine when a decoction of the root was used as drinking water.

Gerard ([1633] 1975:935) noted the slimy property, which, he wrote, helped with urinary stone and also mixed well with oils that mitigate pain. He recommended mallow in "glisters," enemas, for it was of "moderate Heat and moist. It looseth the belly that is bound and is good against the roughness and fretting of the guts, bladder and fundament" (p. 932). He was told by a rabbi that the name *mallow* came from the Hebrew *malluach*.

In northwestern New Spain, cooked mallow was reputed to stop the cough of *tabardillo* immediately. Esteyneffer ([1719] 1978:846) found more than 75 uses for *malva*, including treating paralysis, convulsions, lacrimal fissures, deafness, liver disease, scurvy, *pujos* (abdominal "griping"), hemorrhoids, and urinary problems.

Modern Use. The Mountain Pima cook the entire plant of *M. rotundifolia* to make a drink or a solution for an enema to treat stomach disorders. The Tepehuan treat fever with *malva*. The Tarahumara boil the leaves of *M. parviflora* with a little salt, using the decoction as a drink and to wash the body for fever. They also use tea from the leaves for headache, fever, and gastrointestinal problems and as a poultice for sores. *Malva parviflora*, introduced from the Old World to the Southwest, is decocted for a shampoo but appears to have no medicinal uses.

In Colorado, Mexican Americans make a poultice of the leaves and stems of *M. rotundifolia* for the painful areas of mumps and sore throat. Mexican Americans in southern New Mexico brew the leaves of *M. rotundifolia* for a tea to "wash out the kidneys" and also give this tea with sugar to drink or as an enema for stomachache. They give it as a warm enema for constipation. For *empacho* (infection caused by food stuck in the intestines) the solution should be cool; in addition the abdomen should be rubbed with one tablespoon of olive oil. The tea is used to rinse wounds, abrasions, and infections. The brew also serves as a rinse for hair loss due to cancer treatment. In southern Arizona the decoction, brewed with baking soda and salt, is administered as an enema to reduce the fever of pneumonia. (This is a dangerous use since it can postpone appropriate medical therapy.) Mallow grows plentifully as an undesirable weed in Tucson, but it may also be purchased at a *botanica* and at certain supermarkets.

Phytochemistry. The leaves of *Malva* are high in ascorbic acid and are edible (Duke 1985:291). Its fatty oil contains oleic, palmitic, and stearic acids. Root extracts inhibit the growth of *Mycobacterium tuberculosis*. The flowers contain tannin. All parts of the plant contain abundant mucilage (Font Quer 1979:404). Its emollient effects might explain its popularity.

■ *Mammillaria* (Cactaceae)
PINCUSHION CACTUS

Mammillaria heyderi Muehlenpf.
witculíki, Tarahumara

Mammillaria microcarpa Engelm.
ban mauapi, Pima; *hant iipzx iteja,* Seri

Mammillaria oliviae Orcutt
chollita, Pima Bajo

Mammillaria sheldonii (Britt. & Rose) Back. & Kunth
urimo'o, cabeza de viejita, Pima Bajo

Mammillaria spp.
metzollin, planta que tiene muchas cabezuelas, Aztec; *biznaga,* Spanish; pincushion cactus, fishhook cactus, English

Pincushions are small cacti bearing the spines on crossing spiral rows of teat-like tubercles (from which the generic name *Mammillaria* derives) rather than on ridges. Most of them bear pink to lavender flowers and can be found under shrubs in desert areas. The Spanish common name *biznaga* derives from the Nahuatl *huitzli,* spine, and *nahuac,* around.

Historic Use. The Aztecs used the sap of a species of *Mammillaria* to consume excrescences in the eye. It was given to those who spit blood and also to cure inflammations (Hernández 1959, 2:314).

Mammillaria

1. Used for pain and heart palpitations.

2. Contains alkaloids, which may help to alleviate pain.

3. Contains toxic anhaline.

Modern Use. The Tarahumara use *M. heyderi* to treat earache or deafness. The spines are removed, the plant cut in half and roasted in ashes for four minutes. Then the soft center is squeezed into the ear. The Seri burn off the spines of *M. microcarpa*, then slice and boil the cactus. The resulting liquid is used as drops for earache. The Pima in Arizona treat *M. microcarpa* similarly: the drops are placed warm in the ear for earache or suppurating ears. The Pima Bajo squeeze juice from the leaves of *chollita* on a bit of kapok, which they place in the ear for pain, and toast *cabeza de viejita* for a short time before squeezing to extract the juice, which they rub on the forehead to alleviate headache. A Tucson *curandero* showed me his *cabeza de vieja,* a small *Mammillaria,* saying that it was good for *latido de corazón,* heart palpitations.

Phytochemistry. According to NAPRALERT, *M. heyderi* contains miscellaneous alkaloids. These might explain the use of *Mammillaria* species in earache and headache. *Mammillaria microcarpa* contains the isoquinoline alkaloids including a dopamine, hordenine, and tyramine, which suggest cardiac effects (although treatment for this purpose was reported only by the Tucson *curandero*). Its principal toxin is anhaline.

■ *Mascagnia* (Malpighiaceae)
MASCAGNIA

Mascagnia macroptera (Sessé & Moc. ex DC.) Niedenzu

batenini, babi'ogram, Pima Bajo; *matanene,* Tarahumara; *mantel, gallinita, matanene,* Baja California Sur; *haxz oocmoj* (dog's waist-cord), Seri; *gallinita, mantel rojo,* Spanish; mascagnia, English

Mascagnia is a yellow-flowered shrub or trailing vine native to desert areas in northwestern Mexico. It is little known outside of this area.

Modern Use. The Pima Bajo boil the roots to make a wash for aching limbs. The small leaves are crushed, dampened, formed into small wads, and placed in the ear for earache. The Tarahumara use the plant in tea for malaria and use the bark as a poultice for wounds. In Baja California Sur, a patient with rheumatism may be bathed with a decoction of mascagnia root or branches. The decoction treats wounds and bruises and can serve as a gargle for tonsillitis. An abortion may be provoked with a decoction of *matanene*. The Seri make a tea of the roots to help women gain strength after childbirth. The tea also helps colds and treats diarrhea. I have not heard of its present use by Mexican Americans.

Phytochemistry. According to NAPRALERT, studies have been conducted on *Mascagnia pubiflora* of Brazil, which is generally toxic, causing muscular paralysis. An extract of the branches of *M. macroptera* had 2 + activity against *Staphylococcus aureus, Bacillus subtilis,* and *Candida albicans* (Encarnación and Keer 1991).

Mascagnia

1. Used for pain, colds, wounds, and abortion.

2. Antibiotic activity.

3. Another species caused death in animals.

■ *Matricaria* (Asteraceae)

CHAMOMILE

Matricaria recutita L. Rauchert
[= *Matricaria chamomilla* L.]
manzanilla, Mayo, Paipai, Spanish; chamomile, camomile, German chamomile, English

Chamomile is a small, yellow-flowered herbaceous plant with a scent similar to that of apples (hence the Spanish name *manzanilla,* which translates as "little apple").

Historic Use. *Matricaria* has been known in the Old World through the centuries, especially as a medicine for women's conditions. The botanical name is derived from the Latin *matrix,* womb. It was thought to be 'hot' in the third degree and 'dry' in the second; "it cleanseth, purgeth, or scoureth, openeth and fully performeth all that Bitter things can do" (Gerard [1633] 1975:652).

Introduced to the New World, *Matricaria* was first mentioned in northwestern New Spain by Esteyneffer in 1712. He had multitudinous uses for *manzanilla,* administered in teas, enemas, and fumigations. It was good for stomachache, colic, toothache, ear suppuration, paralysis, constipation, dropsy, scurvy, insomnia, kidney problems, worms, and hemorrhoids. He also recommended it for retained

Matricaria

1. Used as an aid in childbirth, for infant colic, and for pain.

2. Has spasmolytic, anti-inflammatory, and antimicrobial properties.

3. Pollen may cause allergic reactions.

menses, *mal de madre* (uterine disease), retained placenta, and hardened breasts (Esteyneffer [1719] 1978:423, 428, 847).

Modern Use. This may be the most popular herb for Mexicans and Mexican Americans, who use it for a variety of conditions. When a woman begins to have labor pains, she prepares a pot of *té de manzanilla* from the dried flowers. As she drinks the tea, she finds that it may relieve false labor, causing the contractions to cease. If it increases the efficacy of the uterine contractions, making them come more often and harder, she knows she is in good labor. A tea of the flowers or leaves may also be given for infant colic. A tea made by brewing the leaves and adding baking soda and salt is given three times daily for the six weeks after childbirth to "clean out" the mother's system. For *punzadas del ojo*, stabbing pains of the eye, the crumbled flowers are mixed with heated beef fat. This ointment is rubbed on the forehead around the eyes. The patient is told not to go outside or take a bath until the day after this treatment. To make a tea for sleeplessness, add one teaspoon of *manzanilla* to one cup of water and heat until it boils. Remove from heat, cover, and let stand for five minutes before drinking. The Paipai have adopted this introduced plant to treat menstrual problems and *frío*, pathological cold. Mayo women drink the tea during confinement after childbirth, to prevent or treat *frío*. The affected eyes of children are washed with *manzanilla* water.

Phytochemistry. Various constituents in chamomile might explain its panacea-like uses (Duke 1985:297). The oil contains flavonoids and coumarin derivatives responsible for spasmolytic effects (Tyler, Brady, and Robbers 1981:477), which would account for its soothing of colic as well as decreasing uterine cramps. It contains chamazulene and beta-bisabolol. Persons sensitive to ragweeds, asters, and chrysanthemums should avoid this tea, which is prepared from the pollen-rich flower heads (Tyler, Brady, and Robbers 1981). Chamomile contains salicylic and other acids (Font Quer 1979:809). Despite its ubiquitous use, *Matricaria* has been known to cause allergic reactions, although none has been reported since 1973 (Farnsworth 1993a). Potential toxins include borneol, camphor, salicylic acid, and saponin

(Duke 1985:560). *Streptococcus pyogenes* was found to be inhibited by *M. recutita* (Cáceres et al. 1991).

■ *Mentha* (Lamiaceae)

MINT

Mentha arvensis L.
yerba buena, Colorado

Mentha canadensis L.
poleo, Mountain Pima

Mentha piperita L.
menta, Spanish; peppermint, English

Mentha pulegium L.
chichilticxoitl, Aztec; *poleo,* northwestern New Spain; pennyroyal, English

Mentha rotundifolia Huds.
zacatlachichinoa (tlachichinoa herbáceo), Aztec; *maztranzo,* northwestern New Spain; *acete mexicana,* Tepehuan

Mentha spicata L.
yerba buena, Spanish

Mentha sylvestris L.
poleo, Baja California; *yerba buena, hierba buena, póleo,* Spanish; spearmint, English

Mentha viridis L.
tlalatóchietl, poleo chico, Aztec

Mentha spp.
bawéna, Tarahumara

Mints are common, pleasant-smelling herbaceous plants with square stems and opposite leaves. *Mentha* species, which are among the most popular medicinal herbs, are often given the same common names and employed interchangeably, although *M. pulegium* differs pharmacologically in important ways.

Historic Use. *Mentha* species have been known through the centuries in the Old World for treating stomachache. Gerard ([1633] 1975:681) said they were 'hot' and 'dry' in the

Mentha

1. Used for stomachache, menstrual problems, and colic.

2. Antispasmodic and relieves gas.

3. Plants can cause dermatitis; peppermint and spearmint oil is toxic if ingested. Pennyroyal is very toxic.

third degree, quoting Galen and Pliny. Dioscorides differentiated among the mints and *poleo,* noting that the latter was used for women's conditions. Pliny (1938, 6:87–89) stated that mint is good to prevent conception, checks bleeding in both men and women, stays menstruation, heals ulceration and abscess of the womb, and is good for liver complaints, sores on children's heads, tonsils, headache, cholera, etc.

The Aztecs used the decoction of *poleo chico* (*M. viridis*) for flatulence, stomachache, and colic, and as a diuretic and sudorific (Hernández 1959, 2:77). Hernández described it as 'hot' and 'dry' in the fourth degree and perceived it to be more caustic than other kinds of *poleo. Mentha rotundifolia* was used for chills of fever, the fevers themselves, swellings, mange, and leprosy, and was considered 'hot' only to the second degree (Hernández 1959, 2:114).

In northwestern New Spain, Esteyneffer ([1719] 1978: 836) employed *Mentha* for more than thirty conditions, including headache, deafness, paralysis, liver obstruction, melancholy, diarrhea, and dysentery. He recommended the *poleo* varieties for these conditions as well as women's conditions—including problems with menstruation and childbirth—and fevers, gangrene, and poisonous wounds, when administered in a decoction or vapor bath.

Modern Use. The Mountain Pima use *M. canadensis* for sleeplessness, and the Tepehuan use it to alleviate stomach cramps. The Tepehuan use *M. rotundifolia* for earache, which they treat by placing the leaves within the ear. For headache, they drink a tea made from its leaves. The Mayo make an infusion of the leaves of *yerba buena* for stomachache. The Warijio infuse or decoct it for kidney problems and also for sleeplessness. The Tarahumara treat toothache with the leaves, washing the mouth with a decoction or putting fresh or dry leaves on the gums.

Mexicans in Baja California fry the branches of *M. spicata* with *tezo* (*Acacia*) in oil or hen's fat and administer the mixture for cold and cough. For sinusitis, they put two or three drops of the mixture into the nostrils. They take a decoction of the branches for stomachache.

Mexican Americans in Colorado use an infusion of *M. arvensis* as an eye wash and in a bath for fever. As a tea, it is used for stomach ailments and colics. The plant grows in moist areas in the valley and mountains, with plants from the mountains said to be stronger. Mexican Americans in Las Cruces, New Mexico, mix the ground leaves of *M. spicata* with Vick's ointment, applying the ointment around the eye, forehead, and nose and covering the face for an hour to treat eye *punzadas* (a condition characterized by stabbing pains). They brew the leaves as a tea served with sugar to treat diarrhea. A child is also to take bananas and use rice water instead of milk; an adult is given an *atole* (thin gruel) of cornmeal. The tea is also taken for menstrual problems and stomachache and is given to infants for colic. *Poleo* is used to help infants sleep by making the leaves into a pacifier. The leaves are wrapped with sugar in a small cloth that is dipped in warm water. The tea is given for sleeplessness and menstrual cramps.

Phytochemistry. Peppermint contains 1% volatile oil, resin, and tannin, and is a carminative (Tyler, Brady, and Robbers 1981:116). Although many users believe these mints are interchangeable, in fact *M. pulegium* differs chemically from the other (relatively harmless) mints. Abortion is impossible by an ordinary dose of *M. pulegium* in tea, which is safe, but the oil has killed (causing irreversible renal damage) when used as an abortive (Duke 1985:308). The essential oil contains various autoxidation products. The oils of *M. spicata* and *M. piperita* are toxic if taken internally, and the plants of both species can cause dermatitis.

■ *Nicotiana* (Solanaceae)
TOBACCO

Nicotiana attenuata Torr. ex S. Wats.
'*o oB ksaar,* Paipai

Nicotiana glauca R. Graham
sa'usuwam, palo san juan, Pima Bajo; *wipaka,* Tarahumara; *corneton, corneton del monte,* Warijio; *levantate juan, don juan,* Baja California; tree tobacco, English

Nicotiana rustica L.
wipaka, Tarahumara; *punche, punche mexicano,* Spanish; Indian tobacco, English

Nicotiana tabacum L.

pícietl, hierba yetl, Aztec; *tabaco,* Haiti; *tabacum,* henbane
of Peru, Renaissance English

Tree tobacco is a small tree with yellow flowers that is commonly cultivated as an ornamental. The other species listed are herbaceous plants with showy flowers. *Nicotiana* is named after Jean Nicot (1530–1600), who introduced the tobacco plant to France.

Nicotiana

1. Used for pain, for stuffy nose, and to treat an infant's navel.

2. Nervous system stimulant, ultimately leading to respiratory failure, cardiac decompensation, cancer.

3. Contains the alkaloids anabasine and nicotine, highly toxic.

Historic Use. Hernández (1959, 2:80–82) noted that the Haitians called the plant *tabaco* because they used it in their steam baths, which they also called *tabacos.* They transmitted that name not only to the Indians but also to the Spaniards. There were two species, wrote Hernández: both bitter, 'hot', and 'dry' in the fourth degree. They caused drowsiness because the vapors rise to the head. Hernández described the technique of smoking in tubes of paper and inhaling so that the smoke penetrated the chest. This provoked expectoration and alleviated asthma and difficult respiration. It fortified the head, produced sleep, calmed pain, helped the stomach recover its forces, cured migraine, cured inflammations of the spleen, cleaned cancerous wounds, mitigated toothache, healed wounds, calmed painful joints—and on and on. Hernández also spoke of holding tobacco in the cheek, which produced relaxation and restored energy. But he warned that those who have recourse to the help of this plant too often become pale, with a dirty tongue and palpitating throat, suffer liver heat, and die finally attacked by cachexia and hydropsia. This is a fairly good description of terminal cancer.

This information was transmitted by Monardus to the Old World. Gerard ([1633] 1975:357–61), who called the plants *tabaco* or henbane of Peru, noted the numerous conditions cured by this *sana sancta Indorum* (Indian healthy, holy plant). In northwestern New Spain, Esteyneffer ([1719] 1978:868) found all these many uses for *tabaco* including treatment of headache, epilepsy, convulsions, toothache, vomiting, *empacho,* obstruction of the spleen, and whitening of the teeth, as well as to extract insects from the ear.

Modern Use. Many groups use the green leaves of tobacco for headache. The Pima Bajo hold the leaves on the fore-

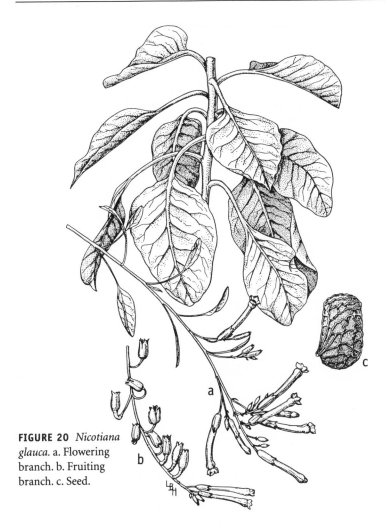

FIGURE 20 *Nicotiana glauca.* a. Flowering branch. b. Fruiting branch. c. Seed.

head with a rag. The Tarahumara apply the leaves directly to the head; the sticky surface of the leaf causes it to adhere. The Warijio coat the leaves with grease to make them stick and apply them to the forehead or to sores. In Baja California, the Paipai smoke the leaves for difficulty in breathing. For headache or rheumatism, the Mexicans in Baja California Sur roast the leaves or bark and apply this to the area in pain.

Mexican Americans in Colorado crush the fresh leaves of *N. rustica* and inhale them to cure stuffy noses. They mix the leaves with sheep fat to treat *dolor de ardo,* heat pain. In New Mexico, *Nicotiana* was put on the navel of the newborn after the cord stump had fallen off. I have not heard of any therapeutic uses of *Nicotiana* by Mexican Americans in Arizona, but *N. attenuata* is employed in such ways by the Hopi, Navajo, Tewa, and Zuni (common names not given).

Phytochemistry. Anabasine and nicotine are the dominant alkaloids in *N. glauca* (Duke 1985:328–29). Its leaves contain rutin. *Nicotiana rustica* is high in nicotine and is classified as a narcotic protoplasmic poison and retardant to neural transmission. The dust of *N. tabacum* is an effective insecticide and molluscicide, but extract of *N. glauca* leaves has no antibiotic activity (Encarnación and Keer 1991).

■ *Ocimum* (Lamiaceae)

BASIL

Ocimum basilicum L.
albahaca, Spanish; garden basil, English

Garden basil is an annual herbaceous plant with oval, serrated leaves and pink or white flowers on spikes; it is cultivated in kitchen gardens where it is protected from high temperatures and freezing.

Ocimum

1. Used for nausea and colic and to decrease breast milk.

2. Has antispasmodic and antibiotic action.

3. Some sweet basil oils contain the carcinogens safrole and estragol.

Historic Use. Gerard ([1633] 1975:674), quoting Dioscorides, noted that basil softens the belly, provokes urine, and dries up milk, which it effects because it is 'hot' in the second degree. Esteyneffer ([1719] 1978) recommended it to be given in steam to stimulate a paralyzed part.

Modern Use. In Baja California Sur, a decoction of basil is used for infant colic and a wash for the eyes. The sap of the leaves is rubbed in the ear for earache. Basil is the fifth most frequently reported medicinal plant in Mexico (Lozoya, Velázquez, and Flores 1988:74), principally as an analgesic and emmenagogue, but also for earache. It is also widely employed by Mexican Americans. In Arizona, the leaves are used in a tea for nausea, emesis, and infant colic, and to decrease breast milk. New Mexicans have used the tea to aid in childbirth and relieve menstrual pains. Basil is believed to have magical properties against evil; it is grown in front of houses and the plant is placed before the shrine of a *yerberia* to protect the store.

Phytochemistry. The essential oil of basil contains cineol, methyl chavicol, estragol, saffrol, and up to 24% linalool (Font Quer 1979:714; Duke 1985:561). An extract of the branches was found to have 2 + activity against *Staphylococcus aureus* (Encarnación and Keer 1991:186).

■ *Opuntia* (Cactaceae)
CHOLLA, PRICKLY PEAR

Prickly pears and chollas are the most common cacti in the southwestern United States and Mexico. Prickly pears have flattened stems ("pads") sometimes bearing fruits along the edges. Chollas have cylindrical stems ("joints"). Both have showy flowers and juicy, edible fruits. Christmas cactus is a small cholla with pencil-thin joints: it should not be confused with the common flat-stemmed house plant *Zygocactus truncatus* (Haw.) Schum., which is also called Christmas cactus.

Cacti in the genus *Opuntia* are commonly divided into two groups: those referred to as cholla and those called prickly pear. Each has specific uses and will therefore be presented separately.

Opuntia acanthocarpa Engelm. & Bigel.
hannam, cholla, Pima

Opuntia bigelovii Engelm.
coote, sea, cholla güera, Seri; *cholla del oso,* Spanish; teddy bear cholla, English

Opuntia fulgida Engelm.
coteext, Seri; jumping cholla, English

Opuntia imbricata (Harv.) DC.
xoconochtli, Aztec; *cholla,* Spanish

Opuntia leptocaulis DC.
hísul, nopal, Mountain Pima; Christmas cactus, English

Opuntia marenae Parsons
xomcahóij, Seri

Opuntia parryi Engelm.
tat kwiss, cholla, Paipai; valley cholla, English

Opuntia–cholla

1. Used for kidney problems and to cast fractures.

2. Some species have shown vasopressor activity from tyramines and cytotoxic activity.

3. No reports of toxicity.

Historic Use. The Aztecs expressed juice from the stem of *O. imbricata* for a remedy that extinguished burning fever and thirst and moistened dry intestines, according to Hernández (1959, 2:312), who declared cholla to be 'cold' in the second degree and 'moist'. Juice was expressed from the fruit and seed and mixed with the juice of *pitahaya*, given to moderate the 'heat' of the kidneys and urine and to counteract bilious and malignant fever. Cholla, said Amarillas (1783) in northwestern New Spain, is washed and boiled to make a plaster cast for fractures.

Modern Use. The Pima gather cholla buds in April, bake them dry, and make a gruel for patients needing a special diet for stomach trouble. The Mountain Pima apply the roasted tender leaves of *O. leptocaulis* to wounds. The Seri utilize many different species of cholla. They boil the core of the root of *coote* for a diuretic tea. They cook the fleshy peel of the fruit with the pulp and seeds for children with diarrhea. For heart disease and heart pain, the spines and skin are removed from a fresh stem and boiled. The resulting liquid is drunk. The inner bark of the root is prepared with leaves of *Argemone* to make a diuretic to clear the urine. The root of *xomcahóij* is cooked in ashes to be eaten for diarrhea. The Apache treated sore eyes by throwing the pith of *O. bigelovii* "on live coals and the smoke is allowed to go into the open eyes" (Hrdlička 1908:234). The most common preparation for diarrhea taken by the Paipai is a tea made from the root combined with an *Eriogonum*. This combats diarrhea by acting as a purge. Mexican Americans in Arizona have used dry cholla stems to cast fractures. In Sonora, Mexico, they use the cholla root for kidney symptoms.

Phytochemistry. According to NAPRALERT, *O. fulgida* and *O. leptocaulis* have alkaloids, while *O. imbricata* contains tyramine and indole alkaloids, alkenes, sesquiterpenes, and monoterpenes.

Opuntia basilaris Engelm. & Bigel.
nopal, Spanish; beavertail, English

Opuntia phaeacantha Engelm.
i-ipai, Pima; prickly pear, English

Opuntia polyacantha Haw.
nopal redondo, Colorado Spanish

Opuntia spp.
nochtli, tuna, Aztec; indian fig, Renaissance English; *iraka,*
ulibecha, Tarahumara; *nopal*, Mayo; *navo*, Yaqui; *laB*,
Paipai; *nopal, tuna simarrona, tuna* (the fruit), Spanish; prickly pear, English

Historic Use. Mexico is the first site of the medical use of
this genus, although the appellation *Opuntia* came from
the Greek name of a plant; Pliny (1855, 3:358) wrote, "In the
vicinity of Opus there grows a plant which is very pleasant
eating to man and the leaf of which, a most singular thing,
gives birth to a root by means of which it reproduces itself."
Despite the description, this cannot possibly be an *Opun-*
tia, which is strictly a New World genus. Oviedo, writing of
the *tuna* in *Sumario de la natural historia de las Indias*
([1526] 1986:156) stated that eating the fruit caused the
urine to turn red. Hernández (1959, 2:311–13) reported that
the Aztecs believed the root, mixed with a species of gera-
nium, would mitigate fever, alleviate hernia, and soothe an
irritated liver. The fruit, eaten with its seeds, prevented diarrhea, especially that caused by 'heat'.

In northwestern New Spain, Esteyneffer ([1719] 1978:
326) reported healing fissures of the palms of hands or soles
of the feet by frying a piece of *nopal* pad and then mashing
it. The skin was first washed with one's own urine or that
from a baby boy or dog, or with a decoction of *malva* or
golondrina. In 1783, Amarillas reported that the people on
the Yaqui River roasted the pads of the *nopal* and extracted
the fluid, which they applied to the side to cure pain there.
His compatriot Pablos (1784), in charge at the Rio Chico,
said the same.

Modern Use. The Pima heat the pads of *O. phaeacantha*
and place them on the breasts of a nursing mother to increase milk flow. The Mayo cook the *nopal* root in water
with cholla root for a tea that is drunk for kidney pain.
They also believe that the water in which the *nopal* root is
cooked will dissolve stones in the appendix. The Tarahumara use several *Opuntia* species. One is employed in a cast
for broken bones, and others are used for the pain of bites

Opuntia–prickly pear

1. Used to treat diabe-
tes, burns, and diabetic
infections.

2. Has antiinflammatory,
estrogenic, hypoglycemic,
and diuretic activity.

3. No reports of toxicity.

and burns. The Yaqui cut up pieces of the stem and put them in a jar with water to drink for diabetes. The Paipai cast a fracture as follows: toast young *nopal* "leaves" until the spines are burned off. Place one on each side, tie tightly, and leave in place until the bone heals. The toasted and sliced "leaf" may be applied to a festering wound. It will also draw out an embedded thorn.

Mexican Americans in Colorado eat the pads of *O. polyacantha* to lower blood sugar. In New Mexico, the pads are diced, soaked in water, and refrigerated, to be taken when thirsty. This, my informant believed, "cures diabetes." For the Arizona recipe, take a 3 by 4 inch piece of the pad, add a small piece of aloe, liquify the pieces in a blender, and stir the resulting liquid into one quart of water. Drink this mixture. For burns, singe off the thorns, open the pad, and apply the cut surface directly to the burn. For a diabetic infection, remove the skin of the tip of the pad, cut it in half vertically, and heat it. Apply the cut surface of the pad to the infected area and cover it with a bandage; remove the pad and repeat three times daily with fresh pads.

Phytochemistry. According to NAPRALERT, many *Opuntia* species have been investigated, with studies conducted all over the world. Flowers, fruits, stems, and entire plants have been studied. *Opuntia ficus-indica* has been found to contain the flavonoids kaempferol, luteolin, and quercetin, as well as rhamnetins and beta-sitosterol. The stem (pad) contains the most active compounds. High in mucilage, this species also has hypoglycemic or antihyperglycemic and diuretic activity. The fruit is high in carbohydrates, especially rhamnose, fructose, galactose, and glucose, as well as vitamin C. The root of *O. polyacantha* has cytotoxic activity.

■ *Perezia* (Asteraceae)
PEREZIA

Perezia hebeclada A. Gray
[= *Acourtia cordata* (Cerv.) B. L. Turner]
pipitzáhuac, zazanaca, coapatli, Aztec; *xararo,* Michoacán;
   pipichaguí, tairago, Ópata

Perezia thurberi A. Gray
[= *Acourtia thurberi* (A. Gray) King & H. Rob.]
ra'mandam, Mountain Pima; *pipichowa,* Tarahumara;
    mata gusano, pipishowa, Warijio

Perezias (the genus is now called *Acourtia* but is referred
to as *Perezia* in every cited reference) are stiff-leaved herbaceous plants up to three feet tall, usually with purple
flowers.

Historic Use. The Aztecs used various species of *Perezia*
(presumed to be *P. hebeclada*) in an enema or as an emetic.
The root, categorized by Hernández (1959, 2:198) as sharp,
bitter, and 'hot' and 'dry' in the third degree, was given in
an enema to calm flatulence and cramps that come from
'cold'. It was thought to evacuate the intestines, eject worms,
and cure diarrhea and *empacho*. However, Sahagún ([1793]
1982:670) said that *pipitzáhuac* root was ground and given
in a drink to those who had "too much interior heat."
Through the purgation of urine, vomit, and stool the interior heat was thought to be mitigated. In Michoacán the
juice of the root of *xararo* was believed to cure the pain of
the "French disease," eject retained semen, and strengthen
the kidneys of women who had just given birth. It was also
given to women to stimulate menses.

The eighteenth-century Ópata drank a decoction of the
Perezia root or took it as a hot enema for stomachache, side
ache, and colic. To make the purge, the decoction was put
out overnight; when the sun rose at dawn, it was to be drunk
until enough evacuation had occurred, at which point
evacuation was to be stopped by taking *atole de maiz,* corn
gruel (Nuñez 1777). The decoction of the root was given to
women with "detained menses."

Modern Use. The Mountain Pima use *P. thurberi* as a violent purge. The Warijio infuse or decoct the roots to treat "a
man with bad penis" or various other genital problems and
also to facilitate menstrual flow. The Tarahumara make a
tea of the root that they take for a purgative and a general
tonic. *Perezia* was a valued purgative for centuries in Mexico and was still listed in the official Mexican pharmacopoeia in 1970 (Lozoya and Lozoya 1982:72). However, today
in Mexico, the root is primarily sold as an abortifacient, or

---

Perezia

1. Used as a purgative and
abortifacient.

2. Has peristaltic and
emetic action.

3. Toxicity has not been
reported, but it contains
toxic compounds.

for "very delayed" menstruation. I elicited knowing looks when I inquired about this medicinal from vendors. Mexican Americans did not volunteer knowledge of the plant.

Phytochemistry. According to NAPRALERT, *Perezia* species that are found in Argentina, Bolivia, Chile, Ecuador, and Mexico and are used as medicines have been studied. The root contains various sesquiterpenes (which have abortifacient action), including sesquiterpene quinones, which are often toxic. *Perezia adnata* increases intestinal peristalsis and emesis (Lozoya and Lozoya 1982:60–79).

■ *Persea* (Lauraceae)

AVOCADO

Persea americana Mill.

ahoacaquáhuitl, Aztec; *gurudusi, laurelillo,* Tepehuan; *aguacate,* Yaqui, Spanish; avocado, English

Avocados are trees commonly cultivated in tropical and semitropical regions. The fruits are the same size and shape as pears but have a leathery skin and one large seed in the center. According to Gerard ([1633] 1975:1606), the botanical name comes from *Persea arbor* (tree of Persia), mentioned by Pliny, Plutarch, Dioscorides, and Galen, via the sixteenth-century botanist Clusius, who did not realize that the avocado was an entirely different tree.

Persea

1. Used in a treatment for rheumatism; also for dermatitis, bruises and sores, and constipation.

2. The seed has antibiotic activity; the leaves, diuretic antihypotensive, smooth muscle relaxant, and uterine stimulant activities.

3. Ground up, the seed is used to poison rats and mice.

Historic Use. The Aztecs, said Hernández (1959, 2:29), used avocado leaves (deemed 'hot' and 'dry' in the second degree) in enemas. The fruit was also 'hot' and thus was thought to excite the sexual appetite and to increase semen. The oil from the seed was said to treat rashes and scars, help dysentery (with its astringency), and prevent hair loss. He described the tree as resembling an oak except that it gave fruit. Sahagún ([1793] 1982) reported that the Aztecs used pulverized avocado seeds for splitting hair, dandruff, and ear ulcers. In northwestern New Spain, Esteyneffer ([1719] 1978) noted that a decoction of the leaf was drunk to dissolve blood clots (p. 267) and was also given as a beverage in place of drinking water during convalescence after *tisis,* spitting blood (p. 270); a decoction of toasted avocado seed was good for diarrhea caused by 'cold' (p. 376).

Modern Use. The Tepehuan crush the seeds of immature fruit to prepare a tea for alleviating diarrhea. They mix the crushed seeds with lard and place this between the cheek and sore gums, or on a goiter. The mashed pulp of an immature fruit is applied to wounds or inflammations. The Yaqui mash the seeds and boil the mashed seed in water to make a tea that they drink to treat diarrhea. To get rid of gray hair, Mayos wash their hair with water in which the seed was cooked. Mexican Americans in Tucson, as elsewhere, soak the seed together with marijuana in alcohol to rub on joints made sore by rheumatism. A recipe for curing dermatitis combines avocado, aloe, and vitamin E cream. The fruit is also valued for constipation because the skin and pulp are seen to be good as cathartics. The seed is always saved because boiled in a tea it may be used as a poultice for bruises or sores. The powdered seed of avocado may be purchased at certain supermarkets in Arizona.

Phytochemistry. Because of the many uses for avocado, beyond those reported above for the American and Mexican West, there have been numerous studies of biological activity and phytochemistry of the seed, fruit, and leaf, as summarized by NAPRALERT. The leaf showed diuretic, antihypertensive, smooth muscle relaxant, and also uterine stimulant activity, as tested on animals. The seed and seed oil contain various steroids as well as vitamins A and D, which might account for treatment of skin disorders, bruises, and inflammations. The seed oil is absorbed quickly through the skin. The seed shows antibiotic activity (Lozoya and Lozoya 1982:26) and is used to poison rats and mice (Morton 1981:236). How a liniment made of avocado, marijuana, and alcohol can treat rheumatic joints is not readily explainable.

■ *Phaseolus* (Fabaceae)

BEAN

Phaseolus acutifolius A. Gray
tepar, Ópata, Tohono O'odham; tepary bean, English

Phaseolus heterophyllus (Humb. & Bonpl.) Willd.
trompillo, Mountain Pima

Phaseolus spp.

Phaseoli Americi purgantes, Renaissance Europe; *quacht-lacalhoaztli, quauhmécatl latifolio, quauhmécatl tenui-folio,* Aztec; *cocolmecate,* northwestern New Spain; *santipús,* Pima Bajo; *cocolmeca,* Mountain Pima; *go-toko,* Tarahumara; *cocolmeca,* Spanish; kidney bean, English

The correct identification of a plant as either *Phaseolus* or *Smilax* has evoked disagreement among botanists since ancient times. *Phaselos* and *Smilax* were both names of Greek beans. The same Spanish common name, *cocolmeca,* is attached to both. There is also disagreement over the etymology of the Spanish name. According to Nentuig ([1764] 1977:62) the name comes from the Ópata *coco,* pain, and *mecca,* away. More likely the name derives from Nahuatl, the language of the Aztecs: *coco,* biting pain, and *mecatl,* vine.

Phaseolus

1. Beans are used as important food; also for pain and sores.

2. High soluble fiber content is antihyperglycemic.

3. Small varieties of lima beans contain a toxin.

Historic Use. Hernández (1959, vol. 2) wrote of many vines and creepers of the Aztecs that were used medicinally, including several that may be *Phaseolus* species. Because of ascribed 'cold' and 'wet' properties, the juice of one instilled in the eyes twice daily could treat inflammation (p. 254), while another treated growths in the eyes (p. 319). The leaves of still another, when mashed and dissolved in water, calmed headache (p. 253).

Gerard ([1633] 1975:1211–16) wrote that the "kindred of the kidney bean are wonderfully many; the difference especially consisteth in the color of the fruit." He wrote of nine kinds of *Phaseolus* (or garden *Smilax,* as Dioscorides called them). Gerard included three kidney beans from America that Clusius had described. All *Phaseolus* species were 'hot' and 'moist', according to "Arabian Physitions," desirable because when eaten they gently loosened the belly and provoked urine without engendering too much wind. (They have a different reputation today.)

Whereas in central Mexico the common name *cocolmeca* refers to a *Smilax* species, in northern Mexico *cocolmeca* is a *Phaseolus* species (Nabhan, Berry, and Weber 1980). Thus the *cocolmecate* described in the eighteenth century by Tamayo (1784) in Arivechi and Nentuig for the

Ópata is likely to be a *Phaseolus*. Tamayo said the root should be cooked to make a purge for "cold in the bones" and also for women with "detained" blood. Nentuig ([1764] 1977) said that the root "restores menstrual blood" and is good for stomachache.

Modern Use. The Mountain Pima make a wash of *P. heterophyllus* by boiling the leaves and stems for fifteen minutes; this may be applied as a lotion to painful arms and legs. The Tarahumara make a tea of the root of *gotoko*, which they drink for pain and also for diarrhea. The Pima Bajo crush the seed of *santipús*, add it to a little grease, and apply the resulting poultice to the lids of sore eyes. Mexican Americans make a decoction of *cocolmeca* to give to infertile women. According to a local *curandero*, the strong purge is believed to cleanse the womb, preparing it for pregnancy.

Today, the beans of *Phaseolus* species are used for food. Tohono O'odham were formerly called Papago (derived from the name *Papavi Kuadam*, tepary eaters), so important were these beans that could grow where there was little water. Scientific studies indicate that the consumption of *Phaseolus* species prevents the rapid rise in blood glucose level after a meal because of their high soluble fiber content (Nabhan 1985:121). When beans were a large part of the O'odham diet, diabetes was not the health problem that it is today.

Phytochemistry. The leaves of *Phaseolus vulgaris* contain—besides various nutrients—allantoin and a hemagglutinin (Duke 1985:362–63). In addition, *P. vulgaris* contains mitogenic lectins that have antitumor properties and cell-mediated cytotoxicity; the seed contains phaseolin, an antibiotic active against fungi (Lewis and Elvin-Lewis 1977: 96–100, 362). Small varieties of *Phaseolus lunatus* beans (lima beans) contain a cyanogenic glucoside (Lewis and Elvin-Lewis 1977:44; Duke 1985:361). The roots "are reported to cause giddiness in human beings and animals" (Duke 1985:363).

■ *Phoradendron, Struthanthus, Loranthus*
(Loranthaceae)

MISTLETOE

Phoradendron bolleanum Eichl.
muérdago, Mountain Pima, Spanish; mistletoe, English

Phoradendron brachystachyum Nutt.
muérdago, Mountain Pima, Spanish

Phoradendron californicum Nutt.
hakvut, Pima; *aaxt, tojí,* Seri; desert mistletoe, English

Phoradendron diguetianum van Tieghem
aaxt, tojí, Seri

Phoradendron spp.
tzavo, Ópata; *kuchóoko,* Tarahumara; *muérdago,* Spanish;
mistletoe, English

Struthanthus diversifolius Standl.
bulikdali, Tepehuan

Loranthus calyculatus DC.
[= *Psitticanthus calyculatus* (DC.) G. Don]
toji, muérdago, Cáhita; *liga, muérdago visco,* Spanish; mistletoe (oak parasite), English

Mistletoes are green plant parasites found in the branches
of trees and shrubs, usually growing only on specific trees.
The generic name *Phoradendron* notes the attachment of
these parasites to trees, from the Greek *phoros,* bearing, and
dendron, tree. The English common name, *mistletoe,* derives from the idea that the plant is propagated by the
droppings of the missel thrush, while the Spanish common
name, *muérdago* (from the Latin *morder,* to bite), reflects
the way the plant is seen to attach to a tree.

Historic Use. In ancient and Renaissance Old World studies of Loranthaceae, the leaves and berries were considered
'hot', 'dry', and biting, and following humoral theory, they
were used to counteract the effects of phlegm that had
"dropped" down inside the body (Font Quer 1979:136–39).
In northwestern New Spain, Esteyneffer ([1719] 1978:747)
recommended a *tóji* decoction for vertigo (p. 176), for *gota
coral,* or epilepsy (p. 181), and in a suppository to evacuate

phlegm from the head and stomach. Nentuig ([1764] 1977: 61) listed *visco o toxi, en ópata tzavo* as one of the herbs already known to medicine.

Modern Use. Mistletoes are employed differently today in the American and Mexican West. The Pima in Arizona boil berries in water until the berries are reduced to a thick mush, then take one cupful as a purge to treat stomachache. The Tohono O'odham use a decoction of the leaves of mistletoe (common name not given) that they find on creosote bushes for stomach and menstrual cramps. The Tepehuan drink a tea of the leaves of *Struthanthus* for difficulties in childbirth. The Mountain Pima eat the berries of two of these mistletoes to alleviate cough. The Tarahumara make a tea from the wood of a mistletoe to treat stomach problems, colds, and venereal disease as well as for a purge. The Seri use *P. californicum* for "illness inside the body." They make *P. diguetianum* stems and leaves into a tea to help women in childbirth and to stem diarrhea. Mexican Americans do not report using mistletoe, but it is available in some supermarkets in Arizona. The bark is recommended for constipation and arteriosclerosis. Juniper mistletoe is used by the Hopi, Navajo, and Zuni.

Phytochemistry. No data have been found to account for the above medicinal uses of *Phoradendron*. These species are used when purgation is desired. (European mistletoe, *Viscum album*, is employed medicinally in Europe and in homeopathy, especially against growth of cancer cells.) *Phoradendron serotinum*, the mistletoe often used at Christmas for decoration, contains beta-phenylethylamine and tyramine, toxic amines that may cause gastroenteritis if eaten in large quantities (Duke 1985:364). (The vasopressor action of tyramine should be kept in perspective—certain foods, including most cheeses, organic meats, and red wine, are high in tyramine; hence, patients taking monoamine oxidase [MAO] inhibitors should not eat these foods although they are safe for others.) *Phoradendron serotinum* is a strong vasoconstrictor and has pronounced ergotlike effects on the uterus (Moore 1979:107). A fatality after a tea made from the berries has been reported, as well as contact dermatitis (Duke 1985:364).

Phoradendron, Struthanthus, Loranthus

1. Mexican Americans do not report using mistletoes today, but other peoples use it for stomachache and women's problems.

2. Purgative and vasoconstrictor actions.

3. Toxic amines could cause gastroenteritis, even fatalities; berries may cause dermatitis.

■ *Physalis* (Solanaceae)

TOMATILLO

Physalis philadelphica Lam.
tómatl, planta de frutos acinosos, miltómatl, Aztec

Physalis pubescens L.
coztómatl, tómatl amarillo, Aztec

Physalis spp.
coronilla, kokovuri, Mountain Pima; *tomatillo,* Tarahumara; *tomate,* Spanish; tomatillo, ground cherry, purple ground cherry, English

Tomatillos are commonly cultivated trailing herbaceous plants native to South America; the fruit is well known today.

Physalis

1. Used for sore throat and cough.

2. Contains niacin and ascorbic acid.

3. Handling plants may cause dermatitis. Unripe fruit contains solanine and other toxic alkaloids.

Historic Use. Hernández (1959, 2:227–32) grouped various genera *(todos los tomates)* in the *Solanum* family (as botanists do today), including those now identified as *Lycopersicon* and *Physalis*. The fruit of the genus now identified from the drawing as *Physalis* he said was 'cold' and 'dry' and somewhat acid; mashed and mixed with chili, it made a sauce that improved the flavor of dishes and stimulated the appetite. The leaves as well as the fruit were thought to be efficacious against Saint Anthony's Fire (erysipelas), lachrymal fistulas, and headache; applied with salt, it was said to resolve mumps. The juice was good against inflammations of the throat, and mixed with lead carbonate, rose oil, and litharge (lead oxide) it was thought to cure ulcers. Instilled in the ear it was said to help earache. Mixed with rose oil, it was considered good for the inflammation of children called *siriasis* when placed on the fontanelle. The tomatillo had traveled from Mexico to the Old World and was described by John Parkinson in 1640 (Hedrick 1972: 433), although this description was missed by Gerard. Esteyneffer ([1719] 1978:293) cited it only to be used in a sauce for stimulating the appetite.

Modern Use. The Mountain Pima use a drink of the leaves of a *Physalis* species for coughing spasms. The Tarahumara treat a sore throat with the fruit. Mexican Americans can buy tomatillos in *yerberias* and grocery stores but are more

likely to use tomatoes *(Lycopersicon).* They squeeze tomatoes, mix the pulp with sugar, and apply this paste on the
sore neck of someone with mumps. Alternatively, they apply slices of the fresh fruit to inflamed areas of the skin.

Phytochemistry. The alkaloid phaseolin is in the fruit
while hygrene is in the roots of *P. pubescens* (Morton 1981:
797–98). The ripe fruit is high in niacin, although the immature fruits are toxic. Toxins in *Physalis* species include
elaidic acid, hydrocyanic acid, and pectin (Duke 1985:563).
The fruit of *P. philadelphica* has been found to inhibit
Staphylococcus aureus, Streptococcus pneumoniae, and *Streptococcus pyogenes.*

■ *Pinus* (Pinaceae)

PINE

Pinus arizonica Engelm. ex Rothr.
úkui, Tepehuan

Pinus ayacahuite Ehrenb. ex Schlecht.
pipíkami úkui (many-spined pine), Tepehuan; *wiyoko,*
   Tarahumara

Pinus chihuahuana Engelm.
sawaka, Tarahumara

Pinus edulis Engelm. (resin of)
trementina, Colorado Spanish; turpentine, English

Pinus engelmannii Carr.
úbisdali, Tepehuan

Pinus spp.
ócotl, Aztec; *ocosaguat,* Ópata; *nnai,* Paipai

Pines are needle-leafed conifers common in the mountains
of the region.

Historic Use. In the Old World, Pliny (1938, 7:25–35) reported various uses for pines and pitches, and these were
repeated by Gerard ([1633] 1975:1356), who assigned the
qualities of 'hot' and 'dry' to the resins and moderate 'hot'
to the cones. He said that *Pinus* and *Picea* (spruce) trees
were good for the chest and lungs, cured the *ptisicke* (now

Pinus

1. Used to treat tuberculosis; also coughing, fever, colds, sores and bruises, wounds.

2. Has antiseptic and expectorant action.

3. Turpentine is potentially toxic taken internally and can cause contact dermatitis.

called tuberculosis), and were good for bladder, urinary, and kidney problems. The pine and its products have been known as medicines through time and space.

Hernández (1959, 2:111) said that there were so many kinds of pines in New Spain that describing them all would be an enormous and useless task. He also reported (pp. 46–47) that the Aztecs used material from *ahoéhoetl* trees (probably *Taxodium macrunatum* Ten., Pinaceae) as medicines for many purposes, including curing ulcers and promoting cicatrization of various skin conditions. He recommended that they be introduced to Spain.

In northwestern New Spain, Esteyneffer ([1719] 1978: 860) reported using the galls of pine in a vapor bath to treat *pujos*, chilblains, herpes, ulcers, dandruff, and buboes. In Arivechi, Sonora, pine was used to make *bilmas*, plasters for fractures (Tamayo 1784). The Ópata woman in labor was placed over a steam bath of pine needles so that "its healthy heat might revive her energies" (Nentuig [1764] 1977:63), a practice also followed by other North American Indians.

Modern Use. The Tepehuan use many pines. They crush the budding cones of *P. arizonica* on a rock and make a tea for alleviating coughing spells or to reduce fever. They apply the gum of *P. ayacahuite* to wounds on the feet, as do the Tarahumara, also using the gum for burns and in tea for colds. The Tepehuan also cook and eat the twigs of *P. engelmannii* for influenza and apply the gum of *Pinus leiophylla* (common name not given) to sores of various kinds. The Paipai place the pitch of a *Pinus* species in a cloth and place this on the chest for coughs, never taking it internally. They mix pine gum with *romero* and another herb to dress bruises.

Alaskan Athabascans use various parts of both pines and spruces for many conditions, including diarrhea and venereal disease, and the gum as a poultice for wounds (Fortuine 1988). The Navajo make a skin salve from pinyon pine pitch and also used it as an emetic; they boiled the leaves with juniper to treat diarrhea (Mayes and Lacey 1989). The Hopi and Zuni use it similarly.

Mexican Americans in Colorado boil *trementina* (turpentine) and mix it with tobacco and honey to make an ointment for removing slivers and treating boils. They mix the resin with powdered *Verbascum thapsus* or a species of *Nicotiana*, honey, and *Anemopsis californica* to make a suppository for treating hemorrhoids. A Tucson *curandero* recommends treating *tisis* (tuberculosis) with pine.

Phytochemistry. The leaf oil of *Pinus elliottii* contains camphene and beta-pinene (Duke 1985:376–77). Pinene is the main constituent of turpentine, and pine resin consists mostly of diterpene resin and acids of abietic and pimaric types. These terpenes have expectorant and antibiotic actions and can cause contact dermatitis. Turpentine is potentially toxic (Foster and Duke 1990:260).

■ *Pithecellobium* (Fabaceae)
MONKEY POD

Pithecellobium
[= *Pithocolobium*]

Pithecellobium confine Standl.
ejotón, palo fierro, Baja California Sur; *heejac,* Seri; *palo fierro,* Spanish

Pithecellobium dulce Benth.
quamóchitl, coacamachalli, quijada de culebra, Aztec; *wamútcali, wamúchili,* Tarahumara; *bacochini,* Mayo; *guamúchil,* Baja California Sur, Spanish

Pithecellobium mexicanum Rose
upatč, palo chino, Pima Bajo; *palo chino,* Spanish

Pithecellobium species are spiny trees with finely divided leaves; they make beautiful ornamentals.

Historic Use. The Aztecs—according to Hernández (1959, 2:265–66), who categorized a plant now identified as *P. dulce* as 'cold' and astringent—used it to treat sores. They boiled the leaves mixed with palm leaves for a decoction to prevent abortion. The leaves, cooked with salt and pepper, cured *empacho*. The juice of the seeds instilled in the nostrils evacuated the "watery humors" of the head. The bark of the root kept contained dysentery and other fluxations. The leaves of another variety, which Hernández (1959,

Pithecellobium

1. Used for symptoms of colds and to counteract pain.

2. Has antibiotic and anti-inflammatory action.

3. No reports of toxicity but bark is rich in tannin.

2:204) called snake's jaw, were used in an ointment that mitigated pain that came from the "Indian plague." Pablos (1784) reported for Rio Chico in northwestern New Spain that *palo chino* bark was used to tan leather and that the gum was burned and powdered to treat sores.

Modern Use. The bark of *palo chino* is crushed and made into a tea by the Pima Bajo to relieve back pains. The Tarahumara boil the leaves of *P. dulce* to make a decoction to wash sore eyes or when vision is clouded. In Baja California Sur, the shoot or bark is fried in oil or hen's fat with a *Mentha* species to make nose drops for colds and sinusitis. *Pithecellobium confine* is an ingredient in a decoction that the Mexicans in Baja California Sur make of the fruit or bark to counteract poisonous stings and treat bruises. They use the shoot or bark of *P. dulce* in a tea to treat stomachache. The Seri make a tea from the whole dry pod of *P. confine* that they drink for cold, cough, and sore throat.

Phytochemistry. According to NAPRALERT, studies of *P. dulce* have showed activity against *Pseudomonas aeruginosa*, sperm, and insects. The leaves contain triterpenes, steroids, and kaempferol. The seed oil contains these as well as beta-amyrin and arginine. The bark contains lupulone. Phytochemical screening has shown that the aerial parts contain flavonoids and tannins. An extract of *P. confine* fruit had 1 + activity against *Staphylococcus aureus* and 3 + against *Candida albicans* (Encarnación and Keer 1991); *P. dulce* branches had 3 + activity against *Staphylococcus aureus*.

■ *Plantago* (Plantaginaceae)
PLANTAIN

Plantago fastigiata Morris
moomsh, Pima; Indian wheat, English

Plantago insularis Eastw.
hataj-en, Seri; woolly plantain, English

Plantago major L.
lantén, llantén, northwestern New Spain; roró, Tarahumara; saxh malk, Paipai; lantén, llantén, Spanish

Plantago mexicana Link
acaxílotl, Aztec

Plantago spp.
  plantaine, Renaissance English; *siñakali, lantén,* Tepe-
    huan; *dud'rum,* Mountain Pima; *yerba del pastor,* Tara-
    humara; *lantén, llantén,* Spanish; psyllium, English

Plantains are small herbaceous plants with leaves surrounding a central stalk of tiny white flowers. These plants
should not be confused with large starchy bananas (*Musa
paradisiaca* L.), which are also called plantains.

Historic Use. In the Old World, Pliny (1938, 7:297–99) recommended the seed, the juice, or the plant itself—boiled,
dried, or powdered and sprinkled in drink—for looseness
of the bowels, while psyllium in water was good for tenesmus. Gerard ([1633] 1975:419–28) summarized the many
uses for this 'cold' and 'dry' medicinal plant, especially for
stopping bleeding and healing sores, using the leaves or
roots. Hernández (1959, 3:224) reported that the root of a
plant similar to what he called *llanten* made a good food
for the Aztecs.

In northwestern New Spain, Esteyneffer ([1719] 1978:
842), like Pliny, recommended a syrup of *Plantago* to treat
liver complaints, worms, and scurvy. He wrote that a decoction of the juice used as a wash or poultice helped exco-
riations of colera (p. 305), mouth and throat sores, lachrymal fissures, and nosebleeds, and helped to fix loose
teeth—he mentioned more than fifty uses. None of these
writers after Pliny recommended using the seed for medicine, as has been the practice of various ethnic groups in
the American and Mexican West.

Modern Use. For diarrhea, the Pima in Arizona mix half a
cup of water with half a cup of ripe seeds of *P. fastigiata*
and allow the mixture to stand. To cure a baby with diarrhea, one tablespoon of ripe seeds is given in half a glass of
water. The Tepehuan crush the roots of a *Plantago* species
on a metate and boil them in a small amount of water for
about ten minutes. The water is strained and drunk (while
warm) for fever. Another *Plantago* species is used for stomach cramps. The Tarahumara use *P. major* in a tea for constipation and another *Plantago* species for gastrointestinal

Plantago

1. Used for constipation,
dysentery, and other gas-
trointestinal ailments.

2. High fiber and mucilage
content acts on intestines.

3. No reports of toxicity.

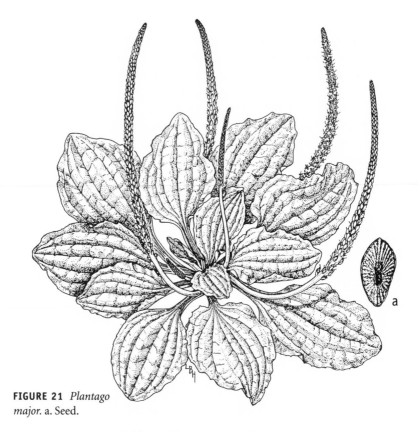

FIGURE 21 *Plantago major.* a. Seed.

problems. The Mountain Pima also use a species of *Plantago* in a tea for diarrhea. The Paipai bind two or three of the large leaves of *P. major* over the forehead to treat headache. The Seri mix whole seeds of *P. insularis* with water, adding sugar if desired, and let the mixture sit for half an hour before it is given. This is considered excellent for children with stomach ailments.

The Navajo use plantain to treat many internal problems—indigestion, stomachache, heartburn, venereal disease, and loss of appetite—and as a diuretic (Mayes and Lacey 1989:82). Alaskans prepare the root of a *Plantago* species as a tonic and use the leaves of *P. major* as a soothing dressing for sore feet (Fortuine 1988). Mexican Americans use the leaves, which grow as weeds near water, to make a potion for dysentery. They may buy *llantén* seeds under various proprietary names in the *botanica,* at certain supermarkets, or at drugstores.

Phytochemistry. The leaves of *Plantago* species contain aucubin, gum, mucilage, resin, and tannin. The seeds contain 18.8% protein, 19% fiber, 10–20% oil, adenine aucubin, choline mucilage, and plantenolic and succinic acid (Duke 1985:386). It seems likely that the high fiber, oil, and mucilage content might explain why *Plantago* is used for constipation and other gastrointestinal problems, even diarrhea, when the ethnic treatment theory requires purging. The tannin in the leaves of *Plantago* may explain its use in treating diarrhea and dysentery. The leaves of *P. major* inhibit *Staphylococcus aureus* (Cáceres et al. 1991).

■ *Plumeria* (Apocynaceae)

FRANGIPANI

Plumeria acutifolia Poir.
cacalosúchil, northwestern New Spain, Spanish; *caguira-guo,* Ópata; *cacalosúchil,* Pima Bajo; *rachiló,* Tarahumara; frangipani, English

Plumeria mollis H.B.K.
cacalosúchil, vipdam us (milky tree), Pima Bajo

Plumeria rubra L.
cacaloxochitl, flor de cuervo, Aztec

Plumeria species are shrubs or trees with alternate leaves, milky sap, and large, showy flowers.

Historic Use. Hernández (1959, 2:268) noted that there were many different kinds of these trees that differed only in the flower. The Aztecs used *Plumeria* as a purgative; the sap of the flower, which "chills and glutinizes," was applied to the chest for pain that came from 'heat'. The medula taken in a dose of two drams he said cleaned the stomach and intestines. In northwestern New Spain, the milk of *cacalosúchil* was taken as a purgative for venereal disease by the Ópata.

Modern Use. The Pima Bajo boil the leaves of *P. acutifolia* to make a lotion for burns and wounds and also moisten slightly the leaves of *P. mollis* for a wound poultice. The Tarahumara use the latex of *P. acutifolia* as a purgative and

Plumeria

1. Used as a purgative and to treat wounds.

2. Has antiinflammatory, antimicrobial, and purgative action.

3. Contains toxic saponin.

for wounds. *Plumeria acutifolia* seems to be absent from the contemporary plant pharmacopoeia of Mexican Americans: perhaps less drastic purgatives are preferred.

Phytochemistry. According to NAPRALERT, *P. rubra* and other *Plumeria* species have been widely studied and used for medicinal purposes where they grow in tropical climates. *Plumeria rubra* has analgesic, antispasmodic, cytotoxic, and hypoglycemic activity. It contains benzenoids; the monoterpenes fulvoplumierin, linalool, and plumericin; the flavonoids kaempferol and quercetin; and saponin.

■ *Populus* (Salicaceae)
COTTONWOOD, POPLAR

Populus angustifolia James
jara, Colorado Spanish; narrowleaf cottonwood, English

Populus dimorpha Brandegee
abaso, alamo, Mayo; *álamo,* Spanish

Populus fremontii S. Wats.
aupa ha hak, alamo blanco, Pima; *xa 'a,* Paipai; *álamo,* Mountain Pima, Spanish; cottonwood, English

Populus tremuloides Michx.
alamillo, Tepehuan, Spanish; quaking aspen, English

Populus spp.
álamo, northwestern New Spain; *vas, torote papelillo,* Pima Bajo; poplar, English

Poplars are common fast-growing, sun-loving trees characteristic of riverbanks and newly forested areas.

Historic Use. Gerard ([1633] 1975:1485–88) described various poplars. Among the medicinal uses for these poplars in the Old World was treating pain, especially from gout and sciatica. He also claimed that the bark, taken with the kidney of a mule, made a woman barren. No tree described by Hernández has been identified as a *Populus.* Esteyneffer ([1719] 1978:216) in northwestern New Spain recommended making a cataplasm from the leaves of a *Populus* species.

Modern Use. Arizona Pima boil a handful of leaves of *P. fremontii* in a pint of water to make a decoction for washing sores. The Pima Bajo place the leaves from a species of *Populus* on the eyelid to treat *mal de ojo* (eye problems). The Tepehuan make a tea from the bark of *P. tremuloides* for alleviating menstrual pains, for stimulating parturition, or as a tonic immediately after childbirth. The Mountain Pima make a tea from the leaves to alleviate cold. They make a tea from the bark of *P. fremontii* that women take in labor; this tea is also thought good for pneumonia. The Mayo cook the bark of *alamo* in water to make a wash for bruises. The Paipai treat wounds with *P. fremontii*.

Alaskan Athabascans make an infusion of the bark of cottonwood as a remedy for stomachache, make an infusion from the buds for colds and internal ailments, inhale the smoke from burning winter branches to soothe cold symptoms, and prepare powdered winter buds with oil for sores, diaper rash, and frostbite. Some boil the leaves to make a treatment for respiratory infections (Fortuine 1988).

Mexican Americans in Colorado find *P. angustifolia* along streams and washes. They chew a fresh twig or leaf to fix loose teeth and treat pyorrhea. A remedy is made from the leaves for venereal disease. In Arizona, for headache, the leaves are boiled and held on the head with a rag, and the water used to boil the leaves is also sprinkled on the head. The leaves may be taken from nearby trees or purchased at certain supermarkets.

Phytochemistry. According to NAPRALERT, studies have been conducted on various *Populus* species, but few that are identified for the American and Mexican West. *Populus nigra*, the European poplar, has been studied extensively and has antibacterial and antifungal activity: it contains apigenin, benzoic acid, and flavonoids including kaempferol, quercitin, and salicin. *Populus tremula* has antipyretic activity from similar compounds. *Populus tremuloides* contains salicin in the leaf and bark and beta-sitosterol in the wood.

Populus

1. Poplar twigs are chewed to treat teeth and gums; leaves are used for headache, pain, respiratory problems, and sores, bruises, and wounds.

2. Most poplars have antipyretic, analgesic, and antibiotic action.

3. No reports of toxicity.

■ *Porophyllum* (Asteraceae)

ODORA

Porophyllum gracile Benth.
heste', siendre, Warijio; *luiB sit*, Paipai; *xtisil, hierba del venado*, Seri; *hierba del venado*, Baja California, Spanish;
odora, English

Porophyllum macrocephalum DC.
chaoacocopin, Aztec

Porophyllum punctatum (Mill.) Blake
hierba del venado, piojo, Spanish

Porophyllum species are herbaceous plants or small shrubs
with yellowish or purplish flowers and a pungent odor.
They grow throughout much of Mexico.

Porophyllum

1. Has been used for colds
and digestive problems.

2. Some species have an-
timicrobial and antiseptic
action.

3. No reports of toxicity.

Historic Use. The Aztecs ate *P. macrocephalum* raw because its flavor was lost by cooking. Hernández (1959,
2:244) found it to be 'hot' and 'dry' in the third degree.

Modern Use. The Warijio crush and infuse the herbage of
P. gracile for colds. Mexicans in Baja California Sur use the
decoction for malaria, pneumonia, stomachache, and diarrhea. They also use it for urinary tract disorders: "cystitis,
urethritis, and kidney calculus." The Seri use the tea for
colds and during difficult delivery: the baby is said to dis-
like the plant's bad odor and is born quickly. A tea is made
from the roots to treat diarrhea and for toothache. The
Paipai make a tea with *P. gracile* that they use for colds and
coughs. Their linguistic relatives, the Havasupai, make a
decoction of the leaves of *P. gracile* as a liniment or tea for
pain, including abdominal pain (Weber and Seaman 1985:
117).

Another species named *hierba del venado (P. punctatum)*, listed among the most commonly reported medicinal plants for Mexico, is used for digestive problems. I had
not heard of the plant from my Mexican American friends,
but it is sold at the local *botanica*.

Phytochemistry. According to NAPRALERT, studies have
found sulfur compounds in *P. gracile*. Other compounds
found in medicinally used *Porophyllum* species include
monoterpenes, thymol, and sesquiterpenes, and in the

aerial parts a triterpene and a flavonoid sakuranetin. In one study no antibiotic activity was found in an extract of *P. gracile* branches (Encarnación and Keer 1991).

■ *Prosopis* (Fabaceae)

MESQUITE

Prosopis glandulosa Torr.
[= *Prosopis juliflora* (Swartz) DC.]
mizquitl, vaina, Aztec; *ku'i,* Pima Bajo; *sako,* Mountain Pima; *juupa,* Mayo; *huupa,* Tarahumara, Yaqui; *nal,* Paipai; *haas,* Seri; *mezquite, péchita, huupa,* Spanish; western honey mesquite, English

Prosopis velutina Woot.
[= *Prosopis juliflora* var. *velutina* (Woot.) Sarg.]
kwi, Pima; *haas,* Seri; velvet mesquite, English

Prosopis spp.
samot, gomilla de Sonora (mesquite gum), Ópata; mesquite, English

Mesquites are thorny trees common in desert areas of the region. Their beanlike seedpods are an important food source for many native peoples. Although these species are all called mesquite in Spanish and English, it is notable that the names in the indigenous languages differ widely from each other. Mesquite is the other panacea of the American and Mexican West, following creosote bush.

Historic Use. Hernández (1959, 3:32) said that there were many species of mesquite, (erroneously) declaring it to be the true *Acacia* of the ancients—which produces gum arabic—and that it was 'cold', 'dry', and astringent. He said that the *chichimecas* made tortillas served as bread from the crushed sweet, edible pods. The Aztecs used the sap of *Prosopis* to treat eye disease. They used a decoction of the bark to repress excessive menstrual bleeding and a discharge after birth that (according to Hernández) if it came from a primipara, a woman who had borne her first child, cured ringworm and favus.

In northwestern New Spain, Esteyneffer ([1719] 1978) recommended cooking the gum for application to joint

Prosopis

1. Mesquite gum is prepared as an eyewash, flour from seedpods as a nourishing food. It is also used to reduce fever, relieve pain, and overcome diarrhea.

2. Mesquite is an excellent source of protein, carbohydrate, calcium, and soluble fiber. It has antibiotic activity.

3. Gum and pollen may cause allergies.

ulcers (p. 678), drinking a beverage made from ground toasted shoots to dissolve kidney stones (p. 405), and powdering the sap for application to bleeding teeth (p. 241). Amarillas (1783) believed that the gum successfully treated eye illness because of its 'hot' quality, while Pablos (1784) used the leaves mixed with *romero* to cure eye problems. The Ópata took a piece of gum the size of a garbanzo, boiled it, then cooled and drank the fluid to cure *mal de madre*. The drink was also given for anxiety and for poisonous stings. The Cochimíes of Baja California mashed the young shoots (common name not given) and expressed the juice to cure sickness of the eyes, which "frequently is suffered in California" (Barco [1768] 1973:64). Nuñez (1777) said that mesquite gave off a gum that was medicinal for many afflictions but especially for the eyes, a treatment that he had used for himself and for his "sons the Indians." Mesquite beans were the most important food for Indians in the Sonoran Desert, including the Yumans, Mohave, Cocopa, Pima, and Tohono O'odham, and were also used by the Yavapai and Pueblo Indians (Niethammer 1974:42). Medicinal use of the gum, leaves, and bark was similarly widespread.

Modern Use. The Mountain Pima soak the bark of *P. glandulosa* for a few hours before boiling the liquid. They take this liquid to reduce fever. They chew or soak the leaves in a pail of water until it is a thick green color. This water is drunk three times daily for *empachos* and diarrhea. The Pima Bajo soak the bark for two days and then boil the liquid to make a purgative. The Pima of Arizona relieve pinkeye with a decoction of *P. velutina:* they pound the green leaves, boil a handful in water, and place them on the eyes. To relieve the soreness of burns, they boil the resin in water and apply the resin. This solution also heals chapped lips and hands. The resin is made into a tea that is held in the mouth to heal painful gums. They make the young roots into a tea to overcome diarrhea. Since mesquite is believed to be a 'cool' plant, it should not be given for a high fever. The neighboring Maricopa pound the gum into powder, mix it with very fine sand, taste it, and if it is not too bitter, sprinkle it on the umbilical stump of the newborn to prevent infection. The Tohono O'odham poultice red-ant

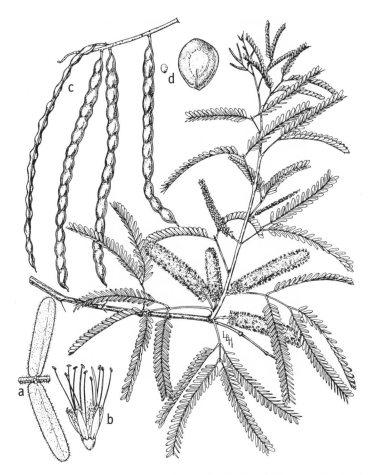

FIGURE 22 *Prosopis glandulosa.* a. Pair of secondary leaflets. b. Flower. c. Seedpods. d. Seeds.

stings with chewed leaves and apply the resin to sores and impetigo.

The Mayo cut the pod of *P. glandulosa* and put it into cool water with lemon, then chill the mixture. They take one cup twice for animal stings. They boil the stems in water or grind the stems into a pulp that they dry, mix with hot water and salt, and then strain. The resulting liquid is drunk once daily on an empty stomach for a purge to treat fever. The Yaqui mash the leaves of *P. glandulosa* into a pulp that they mix with urine and water to make a poultice that they apply to the forehead for headache. For sore throat they pound *huupa* pods into a white powder, heat the powder, and put it on a cloth that is wrapped around the neck. The Tarahumara use the resin in a wash or poultice for the eyes. Although mesquite grows in Warijio and Tepehuan lands, no medicinal use is reported for either group.

The Seri make a drink of the leaves of *P. glandulosa* for an emetic. They dissolve the resin in water to make eyedrops. For a purgative, they take the liquid made from the bark of green or young branches cut into long strips, tied into rolls, and soaked in water. The Paipai boil leaves into a tea to make an eyewash and boil bark into a tea to treat measles. This bark tea acts as a purge and an emetic.

Mexican Americans make the resin of *P. glandulosa* into an eyewash. The resin is sold as a medicinal in certain Arizona supermarkets.

Phytochemistry. Studies of *P. glandulosa* show that the bark and roots contain tannin and alkaloids (Duke 1985: 392–93). Aqueous extracts are antibacterial. Mesquite gum on hydrolysis yields l-arabinose and d-galactose, the high content of arabinose making it an excellent source of sugar. The gum has irritant properties that may cause dermatitis, and the pollen may cause allergic rhinitis, bronchial asthma, or hypersensitivity pneumonitis. No compound is identified that might explain the use of mesquite resin for eye conditions or the leaves as a purgative. Clearly not enough is known about this genus to explain its widespread medicinal uses. The beans are highly nutritious, having protein, carbohydrate, calcium, and soluble fiber.

■ *Psacalium* (Asteraceae)
INDIAN-PLANTAIN

Psacalium decompositum (A. Gray) Rob. & Brett
[= *Cacalia decomposita* A. Gray = *Odontotrichum decompositum* (A. Gray) Rydb.]
péotl xochimilcense, medicina brillante, Aztec; *matariki,* Mountain Pima, Tepehuan, Warijio; *pitcáwi, marariki,* Tarahumara; *maturi, maturín,* Yaqui; *matarique,* Baja California Norte; *matarique, matariki,* Spanish; Indianplantain, English

Matarique is an herbaceous plant with finely divided leaves and a central flowering stalk that bears numerous heads of white flowers. It grows in southern Arizona, Sonora, and Chihuahua.

Historic Use. The Aztecs used a root decoction of *péotl xochimilcense*, judged 'cold' by Hernández (1959, 2:92), "against fever and fluxion of the belly."

Modern Use. *Matarique* is the common name of this plant for speakers of Cáhitan languages, and it is a widely used herb. The tuberous roots are greatly valued by the Warijio in the herbal markets for their medicinal properties. The Tarahumara grind a bundle of roots on a metate and drink with plenty of warm water for a drastic purgative. They pound and boil the roots to make a wash for wounds, and also a drink for cold. The same plant is also used as fish poison. The Mountain Pima drink a brew of the leaves for a sore throat and use this as a wash for cleaning sores. The Tepehuan crush and prepare the roots as a tea for rheumatism.

Matarique has a reputation for healing throughout Mexico. Present-day Mexican Americans and Mexicans use the roots of *matarique* for diabetes (the Tarahumara do not have a similar illness category), as do herbalists in Baja California Norte, who also recommend it for problems of the liver, kidneys, and pancreas. It is harvested by *curanderos* and also may be purchased at certain Arizona supermarkets.

Phytochemistry. Extracts from roots have yielded new sesquiterpenes such as cacalol, cacalone, maturin, maturinin, maturone, maturinone, and dimaturone (Bye 1986: 116–17). In one study, hypoglycemic action was reported, but the active principles were not identified (Pérez et al. 1984). The presence of pyrrolizidine-type alkaloids (Huxtable 1983) can account for its use as a fish poison, but this is ominous for human use. Plants containing pyrrolizidines have caused fatalities.

Psacalium

1. Used for diabetes.
2. Has hypoglycemic action.
3. Contains toxic pyrrolizidine-like alkaloids.

■ *Punica* (Punicaceae)

POMEGRANATE

Punica granatum L.

pomegranat, *malus granata, sive punica,* Renaissance
English; *granada agridulce,* northwestern New Spain;
granada, Pima Bajo, Mountain Pima, Tarahumara,
Mayo, Spanish; pomegranate, English

Pomegranates are common Old World bushes and trees
commonly cultivated as ornamentals because of their beautiful orange blossoms and for their juicy fruits. The Latin
name *Punica* comes from the African *Punic,* the ancient
name for Carthage. The common name *pomegranate* means
many-seeded *(grano)* apple *(pomo).*

Punica

1. Used as a mild
cathartic.

2. Has anthelmintic and
other antibiotic actions
and a high tannin content,
which contributes to as-
tringent action.

3. No reports of toxicity.

Historic Use. Remains of the fruit were found in Egyptian
tombs of 2500 B.C.; it was the symbol of love and fertility in
the Orient (Font Quer 1979:399–401). Dioscorides wrote
extensively of its medicinal uses, as did Pliny. Gerard ([1633]
1975:1451) summarized fourteen uses in the Old World of
the sour pomegranates, which were categorized as 'cold',
'dry', and binding. Pomegranates were introduced to the
New World. In northwestern New Spain, *granada* was employed in a multiple of ways. The juice, decoction, syrup,
mash, roast, and powder were given to take advantage of its
astringent property. Uses included applications for aphth-
ous ulcers, hemorrhage, *cólera,* fever, and liver disorders.

Modern Use. The Pima Bajo boil scrapings of the root for
treating hemorrhoids. They make a tea of the outer portion
of the fruits that they drink for *empacho.* The Mountain
Pima cook the shell, strain it, and drink very hot for diarrhea. The Tarahumara make a poultice with the fruit to relieve painful boils. The Mayo boil the skin of the pomegranate in water to make a tea for sore throat or to apply on
boils. Mexican American infants in Arizona and New Mexico are given a tea made from the rind of the fruit as a mild
cathartic. The rind is sold as a medicinal in certain Arizona
supermarkets.

Phytochemistry. Pomegranate contains 25% tannins and
diverse alkaloids, including pseudopelletierine, pelletier-
ine, and isopelletierine, primarily in the bark of the root,

less abundantly in the trunk and branches (Font Quer 1979: 399–401). These alkaloids are effective vermifuges. The fruit contains glucose, citric acid, malic acid, and a large amount of vitamin C. An extract of *P. granatum* fruit pericarp has been found to be strongly active against *Salmonella typhi* (Pérez and Anesini 1994).

■ *Quercus* (Fagaceae)

OAK

Quercus arizonica Sarg.
rosákame, napako, Tarahumara

Quercus chihuahuensis Trel.
roja, Tarahumara

Quercus crassifolia Humb. & Bonpl.
túái, Tepehuan

Quercus endlichiana Trel.
popuišoli, Tepehuan

Quercus gambelii Nutt.
encino, Spanish

Quercus viminea Trel.
kusi, Mountain Pima; *machichare*, Tarahumara

Quercus spp.
ahoaton, ahoazhoaton, ahoapatli, encino pequeño, Aztec; tu'a, Mountain Pima; *encino*, Spanish; oak, English

Oaks are some of the most common trees at higher elevations throughout the region. Several hundred species occur in Mexico. Their nuts (acorns) are important food sources for many peoples.

Historic Use. Oak is an ancient remedy. Pliny wrote extensively of the medicinal values of oak, and oak galls were found in the excavated remains of the eruption of Mount Vesuvius. Gerard ([1633] 1975:1339–49) also wrote extensively on oak, citing all the Old World ancients from Theophrastus on. The leaves, bark, acorn cups, and acorns themselves, according to Gerard, "do mightily bind and dry in the third degree, being somewhat cold withall" (p. 1341).

Quercus

1. Used for gum problems and other inflammations.

2. Has antibiotic activity and contains tannins.

3. Tannic acid can be cumulatively toxic.

Hernández (1959, 2:15–17) reported on various plants with the prefix *ahoa-* that subsumed oaklike trees. One oak, he said, was astringent, bitter, 'cold', and 'dry'. The Aztecs used it to strengthen postpartum women, fix bones loosened from loins (it was formerly believed that the pelvic bones separated during childbirth), stanch dysentery, and calm the suffering of those fatigued from a long journey, a struggle, or the status of "public official" (Ortiz de Montellano 1990:252). Another oak was said to be 'hot' and 'dry' and was used as an enema to calm problems arising from 'cold', including colic, flatulence, diarrhea, and stomachache.

In northwestern New Spain, Esteyneffer ([1719] 1978: 827) reported the use of galls and leaves of *encino* for various problems: ulcers of the mouth and tongue, loose teeth, nosebleed, hemorrhoids, blood clots, ulcers, wounds, and sores, and for ascites and hydropsia. The acorns *(bellotas)* were used in an *atole* for diarrhea, were powdered for urinary incontinence, and were applied in a poultice to hemorrhoids. The acorns or the buds were decocted and applied in a compress to skin problems, including herpes, weeping sores, and edema.

Modern Use. The Tepehuan boil the bark of a species of *Quercus* to make a refreshing mouthwash. For aching gums, they hold the bark scrapings between the outer gum and the cheek for thirty minutes. The Tarahumara use several *Quercus* species: they make *Q. arizonica* bark into an ointment to treat inflammations and pain, use *Q. chihuahuensis* sap in a tea for heart ailments, and use *Q. viminea* leaves in a tea for gastrointestinal problems. The Mountain Pima also employ the bark of various oaks to treat dysentery, sores (especially on animals), and burns.

Similar uses of *Quercus* are reported for Mexicans. Although not mentioned by any of my Mexican American experts, *Quercus* bark is evidently used since it is stocked in certain supermarkets and herb stores in Tucson.

Phytochemistry. According to NAPRALERT, studies throughout the world have been conducted on some of the hundreds of species of *Quercus*, but not those of the American and Mexican West. Some species show antibacterial, antioxidant, antimutagenic, antifungal, anthelmintic, antiamoebic, antispasmodic, antiviral, and hypoglycemic ac-

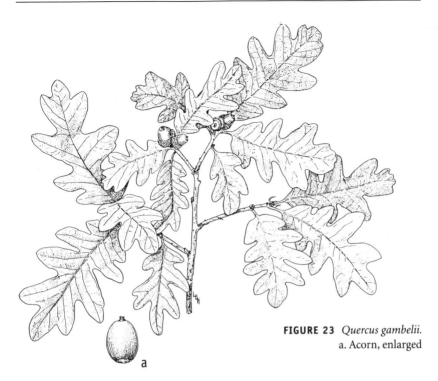

FIGURE 23 *Quercus gambelii.*
a. Acorn, enlarged

tivity. Many compounds have been found in the bark of oak species, with tannins and quercetin perhaps the most significant in explaining ethnic uses. The tannic acid is highly astringent (Font Quer 1979:109–10). The acorns are 50% starch and contain various sugars, fat, and tannin. The galls contain up to 30% gallic and tannic acids, as well as other compounds.

■ *Randia* (Rubiaceae)

RANDIA

Randia echinocarpa Sessé & Moc. ex DC.
papache, Tepehuan, Tarahumara, Mayo, northwestern
   Mexico; *kakáwari*, Tarahumara; *granjel, papache*,
   Spanish

Randia spp.
tempixquitzli, medicina que astringe la boca, Aztec

Papache is a spiny shrub up to ten feet tall, native to dry hillsides of northwestern Mexico, with fruit that darkens on maturity.

Randia

1. Used for urinary tract infections.

2. Contains diuretic and antiinfective compounds.

3. No reports of toxicity.

Historic Use. The plant "medicine that constricts the mouth" reported by Hernández (1959, 2:39) is probably a species of *Randia*. It had, he said, delicious fruit. The Aztecs heated the leaves and applied them to the teeth to strengthen gums and calm toothache. They instilled a decoction in painful nostrils or ears.

Modern Use. The Mayo use a decoction of the bark for boils and drink a tea from the flowers for cough. The Tarahumara and Tepehuan apparently use the fruits only to make a beverage. In Mexico, a tea made from the fruit or leaves is used to treat urinary complaints, as well as coughs, circulation ailments, diabetes, and diarrhea. Mexican Americans have recently become enthusiastic users of *papache* for urinary tract disorders.

Phytochemistry. According to NAPRALERT, other *Randia* species have been used and studied throughout the tropical world, including French Guiana, the West Indies, East and West Africa, Malaysia, India, South Korea, and Thailand. Species investigated contain compounds that have carcinostatic, diuretic, and insecticidal activity. Extensive information is available on the ethnobotany and phytochemistry of *R. echinocarpa* (see Bye et al. 1991): one study found the glucopyranoside arbutin, an effective diuretic and urinary antiinfective, which would make *R. echinocarpa* a reasonable treatment for urinary tract disorders. However, another analysis (Bye et al. 1991) found mannitol, beta-sitosterol, quinovic acid, oxoquinovic acid, ursolic acid, and oleanolic acid; of these, only mannitol is reported to have therapeutic effects on the kidney, but the yield of mannitol is too low when taken orally to be of benefit. Thus there appears to be no scientific support for the ethnic rationale for present-day kidney therapy (Bye et al. 1991:103).

■ *Rhynchosia* (Fabaceae)
ROSARY BEAN

Rhynchosia precatoria DC.
atecuiixtli, ojo de cangrejo (crab's eye), Aztec

Rhynchosia pyramidalis (Lam.) Urban
chanate, sasan, Pima Bajo; *áraiši vuvúji, ojos chanate*

(blackbird's eyes), Tepehuan; *saltipús, sasam wupu'är,*
Mountain Pima; *munísowa,* Tarahumara; *chanate pusi*
(bird eye), Warijio; *negritos, ojo de pajarito* (little bird's
eye), Spanish; rosary bean, English

Rhynchosia spp.
cuchu puusi, ojo de pescado (fish eye), Mayo

Rhynchosia species are twining, beanlike, three-leaved
herbaceous plants. The seedpod contains round, red seeds
with black. Most of the various common names include a
word meaning the eye of some creature, referring to the
appearance of the bean.

Historic Use. Hernández (1960:23) found this plant re-
markable because its pods contained seeds half black and
half red, but he said it had no medical application for the
Aztecs.

Modern Use. The Pima Bajo crush and boil the seeds of *R.
pyramidalis* and strain the liquid to make a lotion for applying to sore eyes. The Tepehuan crush and moisten the
seeds and roots, applying this as a poultice to inflammations. They crush the black and white seeds, add water, and
drink the brew to treat injuries resulting from falls. The
Warijio grind the seeds and mix with oil or grease to apply
as an ointment for sores, bruises, and headache. The Tarahumara make a poultice from the seeds for rheumatism.
The Mountain Pima grind the seeds to powder on a soft
cloth, then dust the eyes with the powder to cure eye diseases. The Mayo take one cup daily of a brew made by
cooking an unidentified part of the plant in one liter of water with cinnamon for cough. Mayo use a species of *Rhynchosia* to increase fertility. I have no information about the
medicinal use of this plant by Mexican Americans.

Phytochemistry. According to NAPRALERT, many species
of *Rhynchosia* have been used medicinally for a variety of
conditions in places including Brazil, the Dominican Re-
public, India, the Ivory Coast, Korea, Malawi, Nepal, Tanzania, Uganda, and Uruguay. Various flavonoids have been
identified in the stem and leaves. Gallic acid was found in
the seeds of *Rhynchosia phaseoloides* and *Rhynchosia min-
ima,* which also contained hydroquinone diacetate. Classed
as a narcotic hallucinogen (Duke 1985:407), *R. pyramidalis*

Rhynchosia

1. Used for sore eyes and
inflammations.

2. Has strong antiinflam-
matory activity.

3. Seed is poisonous and a
narcotic hallucinogen.

was found to contain alkaloids and to have strong anti-inflammatory activity. The attractive but very poisonous seeds have been used in jewelry, with fatal results when consumed by children.

■ *Ricinus* (Euphorbiaceae)

CASTOR BEAN

Ricinus communis L.

kik, biblical Hebrew; *Palma Christi, Ricinus Americanus,* Renaissance English; *apitzalpatli de tehoitzla,* Aztec; *maamsh,* Pima; *Palma Cristi,* Pima Bajo; *mukúkuli,* Tepehuan; *oliráki,* Tarahumara; *higuerilla,* Mayo; *xat mash,* Paipai; *hehe coanj,* Seri; *higuera, higuerilla, Palma Cristi,* Spanish; castor bean plant, English

Castor bean plants are shrubs or large herbaceous plants with large, reddish, lobed leaves. They bear their seeds in a prickly capsule. This plant gets its Latin name from the word for tick, which the seeds are said to resemble. The Spanish name means "little fig." The English common name comes from the beaver (Latin, *castor*), whose oily secretion was used medicinally.

Ricinus

1. Castor oil is widely used as a purgative; the leaves for boils, sores, and pain.

2. Has purgative and antigenic activity.

3. Seeds are deadly poison.

Historic Use. Gerard ([1633] 1975:496–97) wrote that the seed of *Palma Christi* was 'hot' and 'dry' in the third degree and caused vomit and that the oil was good for extreme coldness of the body, as noted by one Rabbi David Chimchi. The Aztec *apitzalpatli* has been tentatively identified as *Ricinus.* Hernández (1959, 2:2), who found the root glutinous and refreshing, said that the Aztecs used it to treat diarrhea and dysentery. This notion would fit the Aztec theory that these conditions require a purgative. In northwestern New Spain, Esteyneffer ([1719] 1978:744) recommended the oil of *higuerilla* seeds as a purgative; the same oil applied to warts would remove them (p. 624). Fomenting engorged breasts with a decoction of the leaves would relieve them (Esteyneffer [1719] 1978:448). The Cochimíes in Baja California used the seed as a purgative, but only for "very robust persons" (Barco [1768] 1973:113).

Modern Use. For headache, the Pima of Arizona shell two to three seeds, eat them, and then tie the brow tightly. The

FIGURE 24 *Ricinus communis.*
a. Leaf. b. Prickly fruit. c. Seed.
d. Double flower cluster (male
above, female below).

seeds are also eaten for constipation. The Pima Bajo crush the fresh leaves by squeezing them, then binding them to the head for headache or applying them to wounds. The Tepehuan apply the leaves as a poultice to goiters and inflammations and crush the seeds for use as a purgative.

The Tarahumara make an ointment for large running sores by mixing leaves with lard before cooking them in ashes wrapped in other leaves. They wrap the leaves of a species of *Ricinus* around the head to treat headache. The Mayo mix the leaf with Vick's ointment, applying this salve to the head for fever. For boils, they take the smallest leaf, on which they put dried, powdered chamomile flower mixed with pork fat. This is wrapped up like a tamale, heated slightly on the *comal* (griddle), and tied onto a boil with a bandage. For headache, they mix ground leaves with pork fat and apply this to the head. The Paipai crush the ripe seed for its oil and place it on boils. The Seri apply mashed seeds to sores on the head. Mexican Americans have used castor oil as a prophylactic purgative, giving it to children "every Saturday." Castor oil poultices are used to treat cancer in alternative medicine.

Phytochemistry. Castor oil consists of ricinoleic acid and other acids, beta-sitosterol, ricinine, and ricin, as well as enzymes. Ingestion of even one seed by a child can be fatal. Ricin produces antigenic or immunizing activity (Duke 1985:408). The seed oil is one of the best purges known, although its taste makes it intolerable to many (Font Quer 1979:188).

■ *Rosa* (Rosaceae)

ROSE

Rosa woodsii Lindl.
rosa, champes, Colorado Spanish

Rosa spp.
rosa de castillo, Paipai; *rosa de castilla,* Spanish; rose, English

Wild roses are common thorny shrubs with beautiful pink or white flowers and juicy red fruits (called "hips"). Roses have been cultivated for thousands of years.

Rosa

1. Rose petals are used for many infant conditions and eye problems.

2. Has some antibacterial, hemagglutinin, and spasmolytic activity.

3. No reports of toxicity but is potentially allergenic.

Historic Use. Gerard ([1633] 1975:1259–71) stated that the various roses have different properties—for example, a red rose is 'cool' and 'dry', and when dried, binding and 'dry', while a white rose is 'cold' and 'moist'. He listed twenty-eight principal uses for white, red, and damask roses. Hernández (1959, 2:228), by contrast, did not list *Rosa* as a medicinal of the Aztecs except for a mention of rose oil as an ingredient in the treatment for the children's condition of *siriasis,* fallen fontanelle (see Kay 1993). In northwestern New Spain, Esteyneffer ([1719] 1978:863–64) cited *Rosa* more than one hundred times, recommending it for back, chest, head, kidney, liver, and lung problems, including colera (heat in the stomach), frenzy, fallen uvula, hemorrhage, melancholy, paralysis, scurvy, syncope (fainting), loose teeth, vomiting, and worms.

Modern Use. Roses are used by few indigenous peoples of the American and Mexican West. An exception is the Paipai, who use roses for fever, menstrual problems, eye infections, and stomachache, as well as to aid birth and as a purge. Since the Paipai have no indigenous name for *Rosa* species, it seems likely that they learned these uses from Mexicans. Mexican Americans in Arizona, New Mexico, and Colorado make a tea of the petals to treat fever. Rose tea also makes a gentle purge for children. Honey may be added, if the tea is for an infant, to improve the flavor (however, biomedicine does not recommend honey for infants because of the possibility of botulism). The tea is drunk for a *tos de calor,* hot cough (that is, acute). To relieve itching, dry rose petals may be ground to a powder and applied. *Chinqual* (rash) may be cured by giving rose flower tea in the infant's bottle. The tea is also given for colic and *torzones de tripas,* intestinal cramps. A Tucson *curandero* makes a solution from the petals for eye problems caused by *calor subido,* intense heat. The tight, dry buds are sold as medicine in certain Arizona supermarkets.

Phytochemistry. According to NAPRALERT, studies have been conducted on rose species throughout the world. The flowers have shown some antibacterial, hemagglutinin, and spasmolytic activity. In Argentina, the flower of *Rosa borboniana* Dep. was found to have antibacterial activity against *Salmonella typhi* (Pérez and Anesini 1994). Studies of the essential oil in the petals show tannins present in the flowers. These compounds may explain ethnic uses of roses.

■ *Rosmarinus* (Lamiaceae)
ROSEMARY

Rosmarinus officinalis L.
romero, Baja California Norte, Spanish; rosemary, English

Rosemary is a small trailing shrub with blue flowers. It is commonly cultivated both as a medicinal and culinary herb and as an ornamental. *"De las virtudes del romero, se puede escribir un libro entero"* ("One can write a whole book on the virtues of rosemary") goes the saying (Font Quer 1979: 652). The Latin name has been traced to its Greek derivation, from words meaning "aromatic bush."

Rosmarinus

1. Used to regulate menstruation, clear vaginal infections, and cause abortion.

2. Antiseptic and abortifacient.

3. Can cause dermatitis.

Historic Use. Gerard ([1633] 1975:1293), summarizing ancient uses in the Old World from Dioscorides through the Arabs, said that "Rosemarie is given against all fluxes of blood" and that the flowers were especially good "for all infirmities of the head and braine proceeding of a 'cold' and 'moist' cause, for they dry the brain, quicken the senses and memorie." Esteyneffer ([1719] 1978:863) in northwestern New Spain also noted these uses and recommended *romero* for epilepsy, paralysis, blindness, cataracts, loose teeth, *mal de madre,* and liver complaints. Rosemary was considered 'hot' and 'dry' in the second degree, according to humoral theory.

Modern Use. In Baja California Norte, *romero* is used for problems of the womb and ovaries, for vaginitis, and to provoke menstruation. It is also used for dyspepsia, stomach and digestive problems, and loss of appetite, as well as for liver problems and rheumatism. The Paipai make a tea from the leaves of another plant (*Trichostema parishii* Vasey) that they call *romero* for menstrual problems and other female complaints.

Contemporary uses by Mexicans and Mexican Americans emphasize *romero* for female problems. In New Mexico and Arizona the leaves are brewed and given as a douche to clean vaginal infection, to regulate menstruation, and also to abort. *Romero* is also placed in bathwater to soothe the skin. Mexicans drink the decoction "to restore health" and to cause abortion. Some refer to *romero* as "the prostitute's herb." Rosemary is also used as a culinary spice; it is available in grocery stores as well as medicinal herb stores.

Phytochemistry. Analyses (Duke 1985:412–13) indicate that the oil from the leaves contains tannins and many monoterpenes, including alpha-thujone. Thujone possesses abortifacient properties. *Romero* also contains expectorants such as alpha-pinene and cineol, camphor, diosmin, trimethylrosmaricinem, luteolin, and apigenin (Winkelman 1986: 120–21). Users of rosemary oil in bath preparations and toiletries should note that it can cause erythema and dermatitis (Duke 1985:413). Recent studies show that rosemary extract is a strong antioxidant (HerbalGram 1993, 28:6).

■ *Ruellia* (Acanthaceae)
RUELLIA

Ruellia californica (Rose) I. M. Johnst.
satóoml, rama parda, Seri

Ruellia peninsularis (Rose) I. M. Johnst.
satóoml, rama parda, Seri

Ruellia tuberosa L.
yerba del toro, Pima Bajo

Ruellias are herbaceous plants with funnel-shaped, hairy, white or richly colored purple flowers. The leaves of *R. californica* are hairy, while those of *R. peninsularis* are shiny and look varnished.

Modern Use. Although this is not a rare genus and is common to Baja California and Sonora, few people report medicinal use of *Ruellia* today. The Pima Bajo crush the leaves of *R. tuberosa* to make a poultice for aching limbs: the poultice is held on for no more than half an hour since it is "very strong." For dizziness, the Seri soak fresh leaves of *R. californica* or *R. peninsularis* in water, then wash the face with resulting yellowish water. For a cold, they boil the bark of the root to make a tea; this tea also treats a stuffy nose. They make the leaves of *satóoml* into an infusion to treat "one who was tired out." The same tea when used as a shampoo cures headaches. Another headache cure is a beverage made by toasting leaves or leafy branches on moderately hot coals, then putting the toasted leaves in a pot with warm water. Use of *Ruellia* is not reported by Mexican Americans.

Phytochemistry. According to NAPRALERT, various *Ruellia* species have been studied, and *R. tuberosa* compounds include apigenin, luteolin, and beta-sitosterol. Alkaloids are present in the shoots. Some *Ruellia* species have demonstrated antibacterial activity.

Ruellia

1. Used for fatigue, colds, and headache.

2. Has antispasmodic, antiinflammatory, and antitussive activity.

3. No reports of toxicity.

■ *Rumex* (Polygonaceae)

DOCK, SORREL

Rumex crispus L.

eviloriva, Tarahumara; *mat kish,* Paipai; *ch'il bikéłł óółitsooígíí* (plant with yellow root), Navajo; *lengua de vaca,* Spanish; curly dock, yellow dock, English

Rumex hymenosepalus Torr.

sivijil, cañaigre, Pima; *kauvati,* Yaqui; *yerba colorada, raíz colorada, canaigre,* Spanish; dock, wild rhubarb, English

Rumex mexicanus Meisn.

axixpatli cóztic, Aztec

Rumex pulcher L.

atlinan redondeado, apatli o remedio que nace junto a las aguas, Aztec

Rumex spp.

bloudwoort, monkes rubarbe, Renaissance English; *lengua de vaca, ruibarbo de los frailes,* northwestern New Spain

Docks and sorrels are common sour-tasting weeds with reddish stems and dry, papery fruits. Perennial, they grow in streambeds. *Axixtli* is Nahuatl for urine, *patli* means medicine, *cóztic* means yellow.

Rumex

1. Used for sore throat, colds, and sores and cuts.

2. Has astringent action from tannins.

3. Large doses can cause diarrhea, nausea, and liver damage; has carcinogenic potential from tannins.

Historic Use. Gerard ([1633] 1975:390) quoted Pliny and Dioscorides as categorizing docks as 'cold' and 'dry'. A decoction of the roots was drunk against diarrhea, dysentery, and nausea. Hernández (1959, 2:6) wrote that the Aztecs used the root and leaves of a yellow *axixpatli.* Now identified as *R. mexicanus,* it is a lacustrine species with a completely yellow root, similar to "friar's rhubarb." *Axixpatli* was thought to provoke urine and evacuate bile. If the root was pulverized and mixed with equal parts of horse, dog, and mouse manure and shells of eggs from which the membranes had been removed, this mixture was thought to break kidney or bladder stones and drive them out of the body. Grinding the root and leaves of *atlinan* (identified as *R. pulcher*) and sprinkling the powder on putrid ulcers was said to cure them, and because of the astringent and 'dry' nature of the plant, would detain diarrheas, dysenteries,

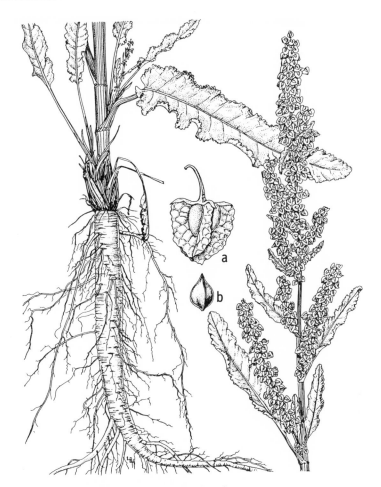

FIGURE 25 *Rumex crispus.* a. Fruiting calyx. b. Achene.

and other fluxations that were 'hot' conditions (Hernández 1959, 2:26). In northwestern New Spain, only Esteyneffer ([1719] 1978:663) mentioned a probable *Rumex* species, recommending that the leaves be placed directly on simple sores and leg ulcers to clean and cure them.

Modern Use. The Pima of Arizona chew the roots of *R. hymenosepalus* to stop cough and treat colds. They make a decoction of the root for washing sores. The Tohono O'odham and Hopi (no common name given) use *R. hymenosepalus* for the same purposes. The Yaqui use *kauvati* for "strep throat," stomachache, and baby teething. The Paipai make a tea from the root of *R. crispus* to treat colds, while the Tarahumara make a tea from the root to treat diarrhea and a poultice from the leaves for sores. Mexican Americans gargle an infusion of the root to soothe a sore throat, although they tell me it tastes terrible. The root is sold as a medicinal in certain Arizona supermarkets.

Rumex crispus is used also by the Navajo: the root and dried leaves are used to treat sores, and a liquid made of the dried leaves is used for sores in the mouth (Mayes and Lacey 1989:33). Alaskans use local *Rumex* species for bladder trouble, stomach trouble, constipation, and diarrhea. The roots are cooked, mashed, mixed with seal oil, and applied to cuts (Fortuine 1988:201–2).

Phytochemistry. The actions of the compounds in *R. crispus* are noted to be contradictory (Duke 1985:414). The species contains anthraquinones, oxalates, chrysophanic acid, emodin, rumicin, and tannins. The astringent action of tannins is utilized in treating sores, a painful throat, and diarrhea; the purgative compounds help with constipation. Tannins are noted to be "effective (but hepatotoxic) when used topically for burns and other exudative conditions, like inflammatory diarrhea. Tannic acid has long been an astringent for diarrhea. The secretory activity and the transudation of fluids in the gut is hindered, and the underlying mucosa may be protected from the effects of irritants in the bowel" (Duke 1985:415). This might explain the Aztec use of dock for urinary complaints. However, overdoses of the root extract may cause diarrhea, nausea, and liver damage. At least one authority (Tyler 1993: 325–26) finds it hard to understand how this simple laxative drug could have retained its ancient reputation for being of value, especially for skin conditions. *Rumex hymenosepalus* contains 18–35% tannins and some anthraquinone (Duke 1985:416). Its high tannin content makes it a potential carcinogen.

■ *Ruta* (Rutaceae)

RUE

Ruta graveolens L., *Ruta chalepensis* L.
ru'ut, Mountain Pima; *ruda*, Tarahumara, Baja California Norte, Spanish; rue, herb grace, English

Rue is a small, delicate herbaceous plant with pale green, thin, highly divided leaves and yellow flowers.

Historic Use. Rue is one of the most ancient of the Old World herbs in common use today. Pliny ([d. A.D. 79] 1855: 252–56), summarizing Hippocrates and Pythagoras, described eighty-four medicinal purposes for this odoriferous plant. It was a principal ingredient in the magical theriac, which was supposed to prevent poisoning. Rue was said to be "productive of fatal results to the foetus." The Romans brought rue to England, where it is grown with other Roman-introduced plants at the Roman Museum in Cirencester. Shakespeare's Ophelia had rue in her bouquet. Gerard ([1633] 1975:1257), who was Shakespeare's botanical advisor, said that rue "bringeth down the sickness [menstruation], expels the dead child and afterbirth" and takes away the pain of an earache; he noted that the cultivated leaves were 'hot' and 'dry' in the latter end of the third degree, with wild rue in the fourth degree, and that contact could be extremely irritating.

In northwestern New Spain, Esteyneffer ([1719] 1978: 864) found many uses for *ruda:* it could be prepared by being fried in oil, infused as a drink, or delivered in steam baths. He recommended the sap for earache and to kill ear worms. He also found it valuable for epilepsy, apoplexy, lachrymal fistula, toothache, stomachache, hydropsy, uterine suffocation, menstrual retention, sciatica, various fevers, inflammations, chilblains, hernias, wounds, sunburn, hypotension, and paralysis. Notably, he recommended rue not only for the bites of rabid dogs but also against witchcraft, a use still admitted to today.

Modern Use. The Mountain Pima use the leaves for a tea that they drink for stomach cramps and painful menstruation. The Tarahumara make a wash from the leaves for earache and a tea from the stem for gastrointestinal problems. Mexicans in Baja California Norte use rue for regulation of menstruation, menstrual cramps, colics, hemorrhages, and nausea and to induce abortion. They also use it for stomach problems associated with hysteria or nervous problems, as well as for the liver, despite its toxicity.

A Mexican woman gave me a shoot of rue to put in my garden. It is a pretty plant but has such an unpleasant odor

Ruta

1. Used for earache and to provoke abortion.

2. Has abortive and spasmolytic action.

3. It is a poisonous abortifacient; handling can cause dermatitis.

that I finally pulled it out. She and other Mexicans and Mexican Americans use this herb in drops for earache. The medication is prepared by gathering the fresh leaves from gardens, drying them (ideally in a clothes dryer that has a nonrotating cycle), then frying the leaves in cooking oil and straining out the leaf remnants. The oil is dropped into the ear canal, then covered with a cotton ball. Alternatively, *ruda* leaves may be rolled in a cigarette paper and smoked; with a paper funnel the smoke is blown into the ear. To treat earache, rue may be made into an ointment with oil or Vick's ointment. For eye pain, this oil may be rubbed around the affected eye. Rue oil appears in over-the-counter ear medication.

Rue is also used for "regulating" menstruation and, as I was told nonchalantly, for provoking abortion. An infusion of the herbage is used as a douche. Mexican Americans brew a tea from *ruda* for menstrual-related pain. For nerves, leaves are cooked with a chocolate bar and taken twice daily. Headache may be treated by rubbing a suspension of the leaves in alcohol on the forehead and the back of the neck. *Ruta graveolens* is sold in certain Arizona supermarkets and also in drugstores that dispense homeopathic remedies.

Phytochemistry. Rue oil contains up to 90% of two ketones, especially methylnonylketone, as well as the alkaloid arborine and furanoquinoline, which are abortive (Duke 1985:417). It is considered a very dangerous abortifacient. Taken internally it is an acro-narcotic poison, and because of its rutin content it is a spasmolytic and thus lowers blood pressure. Handling the plant can cause local irritation such as dermatitis (Font Quer 1979; Duke 1985; Winkelman 1986).

■ *Salix* (Salicaceae)
WILLOW

Salix alba L.
sauce blanco, saúz, jara, Spanish; white willow, English

Salix bonplandiana H.B.K.
sausha ramagara, Mountain Pima

Salix gooddingii Ball
chi ul, Pima; *saúz, sauce,* Spanish; willow, English

Salix lasiolepis Benth.
quetzalhuexotl, Aztec; *awará,* Tarahumara

Willows are trees and shrubs common along riverbanks. The many species are very difficult to tell apart.

Historic Use. In the first century A.D., Dioscorides (1959: 75) wrote of the astringent value of *Salix* as well as its use in pain relief and removal of warts, but he gave no mention of employing it for fever. Pliny (1855, bk. 24, ch. 37) listed treating gout among his many recommendations for the herb. Gerard ([1633] 1975:1388–92) in the Old World categorized the leaves, flowers, seed, and bark of willows as 'cold' and 'dry' in the second degree and astringent, useful for fluxes of blood if the leaves and bark were boiled in wine and drunk. Large amounts of *Salix* pollen were found in coprolites excavated from a southwestern archaeological site dated A.D. 200–800, suggesting possible medicinal use (Reinhard, Hamilton, and Helvy 1991). The Aztecs used *S. lasiolepis* for fever and dysentery (Ortiz de Montellano 1990:252). In northwestern New Spain, Esteyneffer ([1719] 1978) recommended *sauce blanco* for many conditions, including fever, herpes, mange, warts, overabundance of lactation, and also seminal emission, if the leaves were used in a sitz bath.

Modern Use. The Pima make a decoction of the leaves and bark of *S. gooddingii* that they drink for fever; the Tarahumara use *S. lasiolepis* in the same way for that purpose. The Mountain Pima use *S. bonplandiana* bark in a tea to treat fever. Mexican Americans also use an infusion of the leaves and bark for fever and can still purchase the bark for medicine, although aspirin and acetaminophen tablets have largely taken its place.

Phytochemistry. The bark and leaves of *Salix* species contain the glycosides salicine and salicortine as well as tannin (Schauenberg and Paris 1977:107). At one time the bark of willows was used in medicine, but it has since been replaced by synthetic acetylsalicylic acid. Salicylic acid is still used for wart and corn removal.

Salix

1. Used for pain and fever.

2. Has analgesic, antiinflammatory, and antipyretic actions.

3. No reports of toxicity, although aspirin is allergenic.

■ *Salvia* (Lamiaceae)

SAGE

Salvia apiana Jeps.
shalt tai, Paipai; *salvia real*, Spanish; white sage, English

Salvia azurea Lam.
moradilla, Tepehuan; *salvia del monte*, Mountain Pima

Salvia columbariae Benth.
chía, Spanish; chia, English

Salvia officinalis L.
salvia, northwestern New Spain

Salvia pachyphylla Epl. ex Munz.
salvarial de la sierra, Paipai

Salvia tiliaefolia Vahl. Scheich. ex Spreng.
chulisi, Tarahumara

Salvia spp.
chiantzotzolli, planta que se hincha en la humedad, Aztec;
oquisegua (flor mulieris), Ópata; *chía*, northwestern
New Spain; *salvia*, Spanish; sage, English

Hyptis emoryi Torr.
chía, dosábuli, Tarahumara; desert lavender, English

Sages are a diverse group of herbaceous plants, most of which have distinctive scents and showy flowers. The name of the genus *Salvia* and its Spanish common name come from the Latin *salvus* (well, sound), referring to medicinal properties.

Salvia

1. Used as a tonic for nerves, for stomach problems, and as an aid in childbirth.

2. Has antisecretory and carminative actions.

3. No reports of toxicity.

Historic Use. Gerard ([1633] 1975:766) described eight or more *Salvia* species in the Old World, quoting Agrippa and Aetius. He wrote that they called it the Holy-herb because it was considered good for helping women to conceive and to prevent abortion, and also good for the head and brain, for quickening the senses and memory, for palsy, for drawing thin phlegm out of the head, against spitting blood, for cough, for snakebite, for cleaning the blood, and when applied hot, for pain in the side.

The Spanish name *chía* derives from the Aztec name *chiantzotzolli*. Hernández described twenty plants with the *chia-* particle in the name; most have been identified as a *Salvia* or another Lamiaceae. Hernández (1959, 2:69) said that one ounce of the seed of *chiantzotzolli* (*-tlacazolli*, glutinous) taken with water in the morning and in the afternoon protected against fever, dysentery, and other fluxations. The second name that he gave to the plant, *planta que se hincha en la humedad,* translates as "plant that swells with moisture." For the stomach he recommended a plaster made from spiderweb, rose oil, and *chía* seed, to be applied to the belly. *Chía* was considered 'cold' or moderately warm, viscous, and salivous. This seed ground with toasted corn could be stored and when needed mixed with *Agave* syrup (which he said was slightly inferior to our honey) and chili pepper (1959, 2:70).

In northwestern New Spain, Esteyneffer ([1719] 1978) recommended a tea of *chía* to refresh the fevered. He documented many uses of *Salvia,* including for illnesses of the head, epilepsy, paralysis, convulsion, deafness, *tisis,* fainting, and hydrops. The Ópata used a species of *Salvia* to promote menstruation and facilitate childbirth. Pablos in 1784 reported that it was good for women to drink the decoction for a few days after childbirth and that the leaf, carried in a handkerchief, gave protection against bad air.

Modern Use. The Tepehuan make a tea from the flowers and leaves of *S. azurea* to alleviate stomach disorders and also to reduce fever. The Mountain Pima use the leaves in a tea that is drunk before and after childbirth. The Tarahumara make a tea of the whole plant *S. tiliaefolia* for stomach problems and a decoction of the flowers, stem, and leaves of *Hyptis emoryi,* which they call *chía,* for women in childbirth. The Paipai use one *Salvia* to treat stiff neck and colds, and as a childbirth aid, and they employ another *Salvia* in a tea for colds. Mexican Americans buy a combination of leaves that includes a *Salvia* to make a tea for "nerves." *Salvia* is a popular tonic in Tucson, with several species available in certain supermarkets.

Phytochemistry. *Salvia officinalis* produces 2% of an essential oil containing 30% thujone, 15% cineol, camphor, and tannins (Schauenberg and Paris 1977:253). Depending upon the species, *Salvia* contains diterpenes, histamine, and assorted aromatics (Moore 1979:102). The flower and leaf of *S. officinalis* have been found to inhibit *Staphylococcus aureus* and *Streptococcus pyogenes* (Cáceres et al. 1991).

■ ***Sambucus*** (Caprifoliaceae)

ELDER

Sambucus mexicana Presl. ex DC.
[= *Sambucus coerulea* Raf.]
xúmetl, Aztec; *sau, hauku'usi,* Mountain Pima; *tal tal,* Paipai; *saúco,* Spanish

Sambucus spp.
da hap dum, Pima; *saúco,* Pima Bajo, Tepehuan, Spanish; elder, English

Elders are trees and shrubs that generally grow along creek banks. They have many clumps of white flowers that produce juicy red or purple berries.

Sambucus

1. Used for measles, colic, fever, sunburn, infections, etc.

2. Has antiinflammatory, antibacterial, diuretic, diaphoretic, and laxative actions.

3. Stems used as pea shooters have caused cyanide poisoning and severe diarrhea.

Historic Use. *Sambucus* species grow all over the world. The genus provided one of the oldest known medicines in the Old World, and uses for *Sambucus* can be found in an Egyptian medical document, the Papyrus Ebers of 1500 B.C. Gerard ([1633] 1975:1423–24) described eight different *Sambucus* species and cited many uses for the leaves: for removing 'hot' swellings, for those who are burned or scalded, and for those who are bitten by a "mad dog." Gerard quoted Dioscorides as saying that the root cooked in water would heal women's "secret places" if they sat over it. According to humoral theory, *Sambucus* berries are 'hot', 'dry', and purging. The seeds within the berries are thought good for those with dropsy. It has been said that "Elderberry jelly is the panacea of the countryman of Spain" (Font Quer 1979:754); it is a panacea elsewhere, too.

Hernández (1959, 3:218), who identified the Aztec *saúco* as the same tree that grew in Spain, reported that the Aztecs applied the leaves on the forehead and nose for nosebleed, utilized the mashed leaves on sores to cure pain from venereal disease and childbirth, and brewed them into a tea for fever. The people of Michoacán made a decoction of the roots for a purgative they believed would treat diarrhea, reduce fever, and cure other stomach ailments; the sap, injected into the rectum, was thought to clean the intestines. In northwestern New Spain, Esteyneffer ([1719] 1978) recommended the root of the elder tree to reduce swelling of the uvula, as well as for liver problems, paralysis, cavities in teeth, scurvy, scabies, and dandruff.

Modern Use. The Tepehuan make a tea of the flowers of *S. mexicana* for fever and for heart trouble and use the crushed, young leaves in a poultice for cuts. They utilize the flowers of another species of *Sambucus* in a tea for coughing. The Pima Bajo make a lotion to bathe women during childbirth. The Pima make a hot tea with the flowers for a cold, having learned the remedy on their visits to Mexico. The tea is drunk lukewarm for fever. The Maricopa (common name not given) boil the flowers in water and drink the decoction hot for sore throat or stomachache (Curtin 1949). The Mountain Pima prepare a strong drink of *S. mexicana* leaves for fever and stomach disorders and apply the mashed leaves to the head for headache. Mexicans in Baja California Sur make a tea of the flowers for cold and cough, a decoction of the root for rabies. The Paipai use a decoction of the flowers for inflamed eyes, fever, headache, and influenza.

The principal use by Mexican Americans is in a tea for children. Infants are given a tea of the flowers for colic, and children drink it to "bring out the rash" of chicken pox and measles—according to folk belief, if the rash went inward it would cause dangerous sequelae (various cultures hold the same belief). The tea causes sweating, they explain, which will lower fever, too. They also brew the flowers or leaves into a tea for high blood pressure, for whooping cough, for an enema solution to dry up hemorrhoids, as a douche for vaginal infection, or poured over the head for "sun stroke" or sunburn. The berries are used to impart flavor to *atoles,* corn gruels given as convalescent food. In Euro-American culture, the flowers may be fried in pancakes and the berries used for wine.

Phytochemistry. Analysis of *S. mexicana* yields numerous compounds, including sambunigrin, kaempferol, astragalin, and tannins (Ortiz de Montellano 1990). The seeds contain various acids; the leaves, rutin (Duke 1985:423). Children using pea shooters made of elderberry stems can be poisoned by alkaloids and cyanide (Lewis and Elvin-Lewis 1977). An extract of the branches of *S. mexicana* has been found to have 2+ activity against *Staphylococcus aureus* and 1+ against *Bacillus subtilis* (Encarnación and Keer 1991).

■ *Senecio* (Asteraceae)
RAGWORT, GROUNDSEL

Senecio candidissimus E. Greene
[= *Packera candidissima* (Greene) Weber & Löwe]
chucaca, tcukuá, Tarahumara; *lechugilla,* Tepehuan, Spanish; groundsel, ragwort, English

Senecio canicida Sessé & Moc.
itzcuinpatli, veneno de los perros (dog poison), Aztec

Senecio hartwegii Benth.
sopépari, Tarahumara

Senecio salignus DC.
iztacatzóyatl, atzóyatl blanco, Aztec; willow groundsel, English

Senecio vulneraria DC.
calancapatli, iztacpalancapatli, nanahuapatli, palancapatli, Aztec, northwestern New Spain

Senecio spp.
St. James wort, groundsel, ragwort, Renaissance English; *manzanilla de la piedra,* Tepehuan

The genus *Senecio* is a large, heterogeneous group of herbaceous plants and shrubs with yellow, daisylike flowers. Some species have long, elliptical, hairy white leaves, which are also characteristic of many other plants. More than three thousand species are in the genus *Senecio. Packera,* with fifty-one species, has been segregated recently from the genus *Senecio* (Bah, Bye, and Pereda-Miranda 1994); however, in this discussion the original genus name *Senecio* will be used. *Senecio* gets its Latin name from the Latin *senescere,* to grow old, coming from the appearance of flower heads aging prematurely in spring; the English common name *groundsel* comes from the Anglo-Saxon word meaning "pus swallower." The Aztec name *itzcuinpatli* comes from *itzac* (white) and *patli* (medicine).

Historic Use. In the Old World, Gerard ([1633] 1975: 278–81) noted, "The leaves of Groundsell boyled in wine or water, and drunke, healeth the paine and ache of the stomacke that proceedeth of choler. The leaves and floures stamped with a little Hogs grease ceaseth the burning heat of the stones and fundament." Ragwort or St. James wort was noted for being "Hot and dry in the second degree and also cleansing by reason of the bitterness which it hath. It is commended by the later physitions to be good for greene wounds, and old filthy ulcers." Sea ragwort was good for many things: a decoction of the leaves against kidney and bladder stones or against obstructions, especially of the womb, and a bath of the flowers and leaves to bring down menstruation.

Hernández (1959, 2:417) said that *itzcuinpatli (S. canicida)* killed any animal that ate it. However, he heard that it cured leprosy if twelve grains were taken for nine days while the person rested indoors. A plant now identified as *S. salignus,* which Hernández described as 'hot' and 'dry' in the second degree and aromatic, treated fever and chills when the body was anointed with it. He said (1959, 2:64) it had the same uses as absinth *(Artemisia dracunculoides)* or *iztáuhyatl (Artemisia mexicana),* plants that are sometimes mistaken for each other. Sahagún ([1793] 1982:672) reported that *itzcuinpatli* mixed with other materials was good for dandruff, abscesses on the neck, and other infections. *Iztacpalancapatli,* he said, should not be drunk, but the powdered root should be put in putrid sores.

In northwestern New Spain, Esteyneffer ([1719] 1978) found that *calancapatli* prepared as a decoction or powdered and applied was good for ulcers, carbuncles, and genital sores. He also recommended it for cough.

Modern Use. The Tepehuan use a *Senecio* species for a lotion to bathe sick persons and take a tea of the whole plant to relieve kidney discomfort. They use another species of *Senecio* in a tea as a heart stimulant. The Tarahumara are reported to pound the plant of *S. hartwegii* and mix this with a little water, applying it to treat sores and boils. The Tarahumara use *chucaca (S. candidissimus)* as a poultice for

---

Senecio

1. Used for skin infections.

2. Has antiinflammatory, hemostyptic, and antiedemic actions.

3. Taken orally it is poisonous, containing hepatotoxic pyrrolizidines.

sores, especially venereal ulcers. The Navajo use *Senecio douglasii* as a medicine, to be drunk, as a poultice, or in steam for arthritis, rheumatism, and boils (Mayes and Lacey 1989:48). Similar uses of *Senecio multicapitatus* and *Senecio longilobus* are reported for the Zuni and Hopi (Moerman 1985:450).

For some years in southern Arizona, Mexican Americans wanting an herb they called *gordolobo* were supplied with *Senecio* at a local drugstore. At least one fatality and unknown morbidity resulted from ingesting a tea of this toxic plant (Kay 1994). This misidentification was a serious error and might have occurred many times. Recently a specimen from the Tarahumara of some long, elliptical, hairy white leaves from a plant called *chucaca* could not be positively identified at the University of Arizona Herbarium because it included neither a flower nor fruit. In the American and Mexican West, *Senecio* species have been confused with other plants (q.v. *Gnaphalium*). Botanists may be unsure of a plant's identity when a specimen has not been correctly collected. I have no evidence that Mexican Americans knowingly use *Senecio* as medicine.

Phytochemistry. Many species of *Senecio* have been documented as containing pyrrolizidine alkaloids, including *Senecio aureus, Senecio cinerea,* and *Senecio longilobus* (Farnsworth 1993a:36D–E). Although some pyrrolizidine alkaloids (such as senecionine, senecine, and others) have hepatotoxic and carcinogenic effects, others (such as fuchsine and rutin, found in *Senecio nemorensis*) reduce the permeability of capillaries and have hemostyptic effects. These compounds might account for the use of *Senecio* species in treating wounds and skin infections. A large group of these alkaloids manifest atropinelike properties, and some pyrrolizidine alkaloids are classified as deliriants and neurotoxins. Clearly it is dangerous to take *Senecio* species orally.

■ *Simmondsia* (Simmondsiaceae)
JOJOBA

Simmondsia chinensis (Link) C. Schneid.
jojoba, Tohono O'odham; *pnääcol,* Seri; *jojoba, cohobe,* Spanish; deer-nut, jojoba, English

Jojoba is a shrub up to six feet tall, with leathery leaves and acornlike seeds containing an unusual liquid wax.

Historic Use. The first of the sources cited in this study to mention jojoba was Juan de Esteyneffer in 1711, who observed, "The oil of jojobas, which come from Sonora, is a sure remedy for every kind of cancer" ([1719] 1978:600). Also, he noted that it quickly cures any fresh wound: the blood is first pressed out, then a jojoba that has been toasted but not carbonized is scraped until the oil runs out, and dressings are soaked in the black oil and applied warm.

The fame of jojoba became better known to the priests who followed Esteyneffer. An anonymous *receta,* or prescription, dated 1749 (now located in the American collection of the Wellcome Institute) lists eight "virtues" that were later summarized by Barco ([1768] 1973). According this anonymous report to the Medical Tribunal of Mexico, jojobas were good for urinary retention caused by an excess of phlegm (to be treated with five or six nuts dissolved in wine, broth, or hot water); for gas, indigestion, and other ills that originate from those conditions, such as stomachache, headache, obstructions of the stomach or the belly, bowel irregularity, and any indisposition of the stomach that comes from 'cold' (to be treated with seeds dissolved in wine, broth, or hot water or simply cooked and eaten); for acid indigestion, cramps in the gut, or nausea that comes with hunger (to be treated with seeds taken straight or dissolved in wine or broth for a few days); for wounds (to be treated with a balsam made from a few jojoba nuts toasted and rubbed until the dark oil has dissolved, then applied with a feather as many days as necessary) or old wounds and spasmed sores that do not drain, are a bad color, and are swollen (to be anointed with the hot oil and then covered); for stopping cancer which is beginning and curing that which has already set in (by expressing the oil and anointing the cancerous part with it until the cancer is eradicated); for giving birth (by having the woman in labor drink five or six nuts dissolved in broth or wine when the urge to push comes); or for any kind of fever, especially tertian fever.

Segesser (Treutlein 1945:188), reporting to his provincial superior in the 1730s, noted, "There is a large, tree-like

Simmondsia

1. Used in shampoos and face creams; also for sores.

2. Has carcinostatic and bile-stimulating actions.

3. No reports of toxicity.

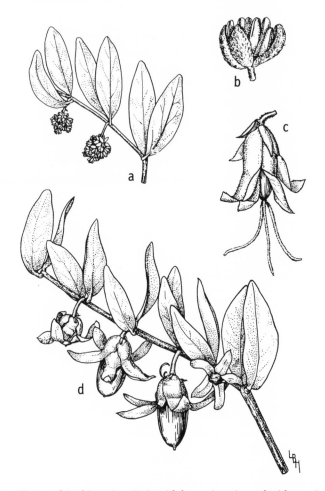

FIGURE 26 *Simmondsia chinensis.* a. Twig with leaves in pairs and with staminate inflorescences. b. Staminate flower. c. Pistillate flower. d. Branch with fruits.

shrub which bears fruit similar to the acorn. It is called *co-hobe* fruit, tastes like a hazelnut, and is very oily. Oil pressed from it is highly prized. The Fathers Procurators in Rome made very urgent request for *cohobe* because of its very high value in Rome. I do not doubt that physicians and apothecaries in Germany know this fruit, though perhaps by another name."

Nentuig ([1764] 1977:46–47) said, "Jojoba is a fruit well-known by this name, by which it is called by all the natives of Pima and Opata.... If they are wounded, and the wound is in such a place that it allows the arrow to be drawn out, they place in the wound one, two or more jojobas, indeed as many as there is room for. In this manner swelling is prevented." He also recommended jojoba for intense belly ache.

Pfefferkorn (Treutlein 1940:65) likewise spoke of jojoba's effects on the stomach: "Since it is pleasant to eat, it is the more popular as a mild and good remedy for stomach aches, being especially helpful in cases where the stomach has been chilled. It must be taken rather sparingly, however, because it is Hot and too much is constipating." Barco ([1768] 1973) had said that before the conversion of the (Baja) California Indians, they did not know the "healing virtues" of jojoba despite the ubiquity of the plant. Afterward, however, the plant's medicinal uses were forgotten.

Modern Use. The Seri use jojoba: "The nuts were placed in hot ashes, removed before they burn and then crushed on a metate. The oil and blackened portions were applied to sores on the head. To relieve eye soreness, the seeds were ground, put into a cloth, and the liquid squeezed into the eyes. Raw green seeds were chewed to deaden the pain of a sore throat. As a remedy for a cold, the seeds were crushed and cooked in water for a short time. The same liquid was taken near the time of birth to aid in delivery" (Felger and Moser 1985:365). The Tohono O'odham parch and pulverize the nuts (common name not given) and apply the dry powder to sores.

Jojoba is now grown for industry. The oil is an ingredient in face creams and shampoos. It is also used in rockets, where its resistance to high temperatures is a valuable substitute for the oil of the endangered sperm whale. Mexican Americans did not report using jojoba, but the seeds are now sold for medicine in certain Arizona supermarkets.

Phytochemistry. Jojoba contains simmondsin, isorhamnetin-3,7-dirhamnoside (carcinostatic), and rutinoside, which is bile stimulating (Duke 1992:560–61). Perhaps some of the extravagant claims made in the eighteenth century have a chemical basis.

■ **Smilax** (Liliaceae)

SARSAPARILLA

Smilax aristolochiaefolia Mill.
[= *Smilax medica* Schlecht & Cham.]
mecapatli, zarzaparrilla, Aztec; *zarzaparilla,* Spanish; sarsaparilla, sasparilla, English

Smilax spp.
salsaparilla, rough binde-weed of Peru, *smilax aspera,* common bind weed, china root, Renaissance English; *cozolmécatl, cuerda de cuna,* Aztec

Smilax is a slender vine with glossy foliage. The confusion between *Smilax* species and *Phaseolus* species when identifying *cocolmeca* is mentioned under the latter entry. The common name in Spanish derives from *zarza-* (thorn) and *-parilla* (vine). The Aztec name, from *cozol* (cradle) and *mecatl* (vine), indicates its use.

Smilax

1. Used to flavor drinks.

2. Diuretic and sudorific.

3. Contains saponins, which in unusually large doses could be harmful.

Historic Use. Hernández described several *mecapatli,* including plants now classified as *Ipomoea* and others of the family Convolvulaceae. The first, "which the Spanish call *zarzaparilla,*" was extolled by Hernández (1959, 2:248–49). He believed this plant to be the same *smilax aspera* described by Dioscorides (therefore, Hernández said, he did not bother to describe its form). However, Dioscorides said nothing about its "temperament" or "virtues" of provoking sweat, calming joint pain, and conquering incurable diseases, omissions Hernández filled. A concoction of *zarzaparilla* that was enthusiastically described by Monardes was further endorsed by Ximénez ([1615] 1888:229) for every kind of illness. Ximénez had also heard of marvelous cures from *cozolmécatl,* especially for ulcers caused by the "French disease" (syphilis). The Aztecs used it effectively for joint pain because of its diaphoretic and diuretic actions (Ortiz de Montellano 1990:181). Gerard ([1633] 1975:861) said "the roots are of temperature Hot and Dry, and of thin and subtil parts, insomuch as their decoction doth very easily procure sweat. The roots are a remedie against long continual paine of the joynts and head, and against cold diseases." He complained that Monardes wrote no description of the plant, saying only that the root was long. He also grouped many Convolvulaceae into his category of bind-weeds. In northwestern New Spain, Esteyneffer ([1719] 1978) recommended powder of *zarzaparilla* in a mixture for chronic cough, in an ointment for "gallic" (venereal) ulcers, and in a syrup for the disease.

Modern Use. Today *zarzaparilla* as a purgative is sold in the various Mexican and Mexican American medicinal herb stores that I visited, but it has been superseded as a cure for venereal disease. It is sold in certain supermarkets. Sarsaparilla is widely used as a flavoring in a soft drink.

Phytochemistry. Several species of *Smilax* contain the steroids sarsasapogenin, smilagenin, sitosterol, stigmasterol, and pillinasterol, and saponin glycosides. Sarsaparilla is approved for food use, but in unusually large doses the saponins could possibly be harmful (Duke 1985:446).

■ *Solanum* (Solanaceae)
NIGHTSHADE, HORSENETTLE, POTATO

Solanum americanum Mill.
chichiquelite, Mountain Pima; *chichikalite,* Tarahumara

Solanum eleagnifolium Cav.
[= *Solanum hindsianum* Benth.]
huitztomatzin, tómatl espinoso (spiny tomato), Aztec; *vakwa hai,* Pima; *mariola,* Baja California Sur; *hap itapxén* (mule-deer; its inner canthus), Seri; *trompillo,* Spanish; white horsenettle, English

Solanum erianthum D. Don
malabar, Baja California Sur

Solanum madrense Fern.
saca manteca, Tepehuan

Solanum nigrum L.
toonchichi, planta que da frutos acinosos amargos, Aztec; *hierba mora, chichiquelite,* Tepehuan, Baja California Sur, Spanish; nightshade, English

Solanum nodiflorum Jacq.
laltu'uskul, Mountain Pima

Solanum rostratum Dun.
soíwari, Tarahumara; *baachipori,* Mountain Pima; buffalo-bur, English

Solanum tuberosum L.
papa, Spanish; potato, English

Solanum verbascifolium L.
hikuli, hukuli, saca manteca, Tarahumara; *wahtauwi, corneton del monte,* Warijio

Solanum spp.
nightshade, Renaissance English

The genus *Solanum* contains more than a thousand species, including such important food plants as the potato and the eggplant. Nightshade is a common trailing weedy herbaceous plant with purple flowers; the purplish black fruits are edible in some geographic regions of North America but poisonous in others. *Solanum nodiflorum* and *S. americanum* are very similar, while *S. erianthum* and *S. verbascifolium* are somewhat larger, with yellowish hair on the leaves. Buffalo-bur, white horsenettle, and *S. madrense* are prickly, herbaceous weeds.

Solanum

1. Horsenettle tea was once used for diabetes; raw potato for fever.

2. Some species have antibacterial and antiinflammatory action.

3. Solanine, found in green parts of fruit, can cause poisoning.

Historic Use. Gerard ([1633] 1975:339–41), quoting Old World writings, noted that *S. nigrum* is used for those infirmities that have need of cooling and binding, qualities it has in the second degree, and that it should be mixed with rose oil and applied to the heads of children against the condition of *siriasis,* and to treat St. Anthony's fire, lachrymal fistulas, and other 'hot' diseases. Hippocrates had warned of its dangers, however.

Hernández (1959, 2:230) said that the Aztecs mashed the bark of the root of *huitztomatzin* (*huitz-,* spiny; *toma-,* tomato; *tzin-,* small) for a medicine to treat a fever. Because it was bitter, 'sharp', and 'hot' almost in the third degree, it "evacuated all humors by the inferior route." A species of *Solanum* that gives acidic, bitter (-*chichi*) fruit and that Hernández (1959, 2:231) thought was like the plant called belladonna was used in a decoction to clean the stomach, while the juice was good for the eyes and to resolve tumors.

In northwestern New Spain, Esteyneffer ([1719] 1978) found numerous uses for *hierba mora (Solanum nigrum),* a

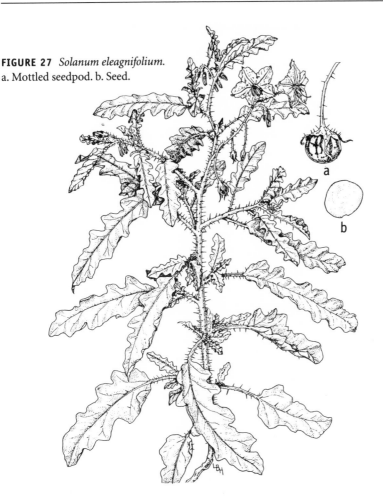

FIGURE 27 *Solanum eleagnifolium.*
a. Mottled seedpod. b. Seed.

'cold' remedy to counter 'hot' diseases: breast inflammations after childbirth, toothache, sciatic pain, and phleg-
mon, and also for fomentations on cancerous parts, genital
ulcers, and liver disorder. Amarillas writing in 1783 said that
the Yaqui cooked the leaves of *chichiquelite* in water and
drank the water to treat *tabardillo* (typhus).

Modern Use. The Mountain Pima treat fever with a tea of
S. americanum and measles with a tea made from the leaves
of *S. nodiflorum.* The Pima crush the dry yellow berries of
S. eleagnifolium and hold them to the nose. This causes violent sneezing and serves as treatment for a cold. The Tepehuan use the leaves of *S. madrense* to make a lotion for
scorpion stings. They soak the leaves of *S. nigrum* to make

a lotion for sores from worms and make a strong tea of these leaves for intestinal worms. The Tarahumara heat the leaves of *S. verbascifolium* in ashes and apply them to wounds or sores; the Warijio coat the leaves with grease and apply them to the forehead for headache or on sores. The Tarahumara use the leaves of *S. americanum* for treating sores and wounds and for rheumatism and drink the leaves of *S. rostratum* in tea to help relieve menstrual pain.

In Baja California Sur, the branches of *S. erianthum* are cooked into a tea to treat blood pressure, stomach problems, and colds. A decoction of the root or flower of *S. eleagnifolium* mixed with rose and sugarcane pulp is given to fortify the uterus of the woman in childbirth, promote menstruation, expel the placenta, and treat earache. The Seri use the tea as a remedy for diarrhea. The Baja Californians Sur roast *S. nigrum* leaves to make a poultice with milk, to be applied to the area infected in erysipelas. The leaves of *S. nigrum* are sold as a medicinal in certain Arizona supermarkets.

Mexicans and Mexican Americans used to treat fever by slicing a fresh potato and placing the slices on the head, around the neck, or on another site. The potato "would turn brown from the heat, reducing the fever." Another method called for dipping the peeled potato slices in vinegar and placing them on the forehead and soles of the feet. Many adults remember this treatment. For diabetes they made a tea from the leaves of *S. eleagnifolium*. Neither of these treatments is currently practiced.

Phytochemistry. An extract of the branches of *S. erianthum* has been found to have 2+ activity against *Staphylococcus aureus; S. eleagnifolium* to have 2+ activity against *Staphylococcus aureus* and *Bacillus subtilis;* and *S. nigrum* to have 3+ activity against *Staphylococcus aureus* and 2+ against *Bacillus subtilis* (Encarnación and Keer 1991). Treatments for skin infections and warts have used *Solanum dulcamara* at least since A.D. 150. The seeds of *S. nigrum* contain linoleic, oleic, palmitic, and stearic fatty acids, sitosterol, and cholesterol (Duke 1985:450). Fruits contain diosgenin and tigogenin. Roots, shoots, and mature fruits are low in alkaloids, but green fruits contain solanine and other alkaloids, saponins, and tannin. Symptoms of solanine poisoning include diarrhea, mydriasis (dilation of the pupil), panic, excitation, coma, hyperthermia, paralysis, and rarely, fatality. Green potato parts also contain solanine, which can cause poisoning. Leaves of both *S. americanum* and *Solanum nigrescens* have been found to inhibit *Staphylococcus aureus* (Cáceres et al. 1991).

■ *Sphaeralcea* (Malvaceae)
GLOBE MALLOW

Sphaeralcea ambigua A. Gray
desert hollyhock, Paipai; *jcoa*, Seri; *mal de ojos*, Spanish; desert mallow, desert hollyhock, globe mallow, sore-eye-poppy, English

Sphaeralcea angustifolia (Cav.) G. Don
tlaltzacutli, Aztec

Sphaeralcea coccinea (Nutt.) Rydb.
yerba de la negrita, Spanish

Sphaeralcea emoryi Torr. ex A. Gray
hadamdak, Pima

Globe mallows are common desert herbaceous plants with bright orange flowers. The Latin and English common names of *Sphaeralcea* come from the spherical shape of its fruits, while the Spanish common name comes from its reputation for treating eye conditions. The Aztec name comes from *tlal-*, wild, and *cutli*, gummy.

Historic Use. The Aztecs pulverized the leaves to make a dose for treating diarrhea. Hernández (1959, 2:120) believed that since it was glutinous, 'cold', and malvaceous in nature, it would take away hives and detain other fluxations.

Modern Use. The Pima pound the root of *S. emoryi* and boil it in a little water to make a drink for diarrhea. The Paipai boil the leaves of *S. ambigua* to make a wash to treat pimples on a child's hands and face. The Seri remove the bark from the roots of *S. ambigua* by pounding them on a metate; the resulting frothy and slimy pulp is made into a tea thought to be good for diarrhea and sore throat, or to be used as eyedrops.

Sphaeralcea

1. Used to treat boils and infected cuts.

2. Has emollient properties.

3. No reports of toxicity.

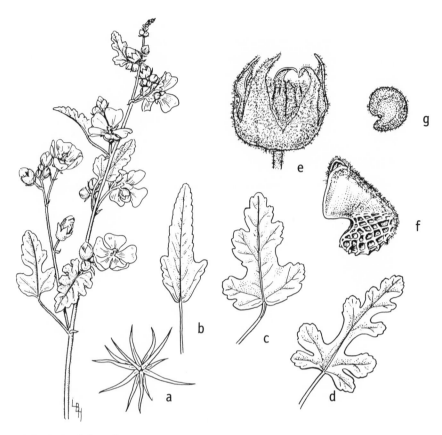

FIGURE 28 *Sphaeralcea emoryi* var. *variabilis.* a. Starshaped hair. b, c, d. Typical leaves showing variation in shape. e. Fruiting calyx. f. One section of seedpod. g. Seed.

Mexican Americans in Colorado crush the leaves and the orange flowers, mix these with sugar, and apply the mixture to boils. The crushed herb is applied to infected cuts and cracked hands. They make an infusion of leaves and flowers for a rinse to prevent hair from falling out. Mexican Americans in Arizona can purchase the herb in certain supermarkets. The Hopi, Navajo, and Tewa also employ *Sphaeralcea* species as medicine (Moerman 1985: 465–66).

Phytochemistry. Little is known about the biological activity or compounds contained in *Sphaeralcea*. According to NAPRALERT, phytochemical screening has revealed the presence of alkaloids. Like other Malvaceae, *Sphaeralcea* has emollient properties, and its root may be used to substitute for *Althea officinalis* L. (Martínez 1969:433), for example, to treat skin infections and injuries (Lietava 1992:365).

■ *Tagetes* (Asteraceae)

MARIGOLD

Tagetes elongata Willd.
San Paulillo, Mountain Pima

Tagetes filifolia Lag.
anisillo, Warijio

Tagetes lucida Cav.
yyauhtli, hierba de nubes, Aztec; *hierba de Santa Maria,* northwestern New Spain; *guamusi, anis,* Ópata; *yerba anís,* Mountain Pima; *basigó, coronilla,* Tarahumara; *yerba anís, Santa Maria,* Warijio, Spanish

Tagetes micrantha Cav.
anisillo, Mountain Pima; *yeyésowa,* Tarahumara

Tagetes spp.
cempoalxóchitl, flor de veinte hojas, Aztec; *yerba del venado,* Baja California; marigold, English

Wild members of the genus *Tagetes* are similar to the cultivated marigold (*Tagetes erecta* L.) but smaller and generally anise-scented. The genus *Tagetes* is wholly American, although it was named after Tages, an Etruscan deity and the grandson of Jupiter, who sprang from the ploughed earth (Coombes 1985:190). It has been introduced throughout the world, mostly as an ornamental. The common name *marigold* is given to various *Tagetes* species and to *Calendula,* another Asteraceae.

Historic Use. *Yyauhtli* was highly important to the Aztecs. Used for fever, fear, dementia, and the aftereffect of lightning, it was also believed to stimulate menses, provoke abortion, expel the dead fetus, increase milk, serve as an aphrodisiac, cure a child's *empacho,* heat the stomach, thin humors, dissolve tumors, cataracts, and stones in the kidneys and bladder from "accumulated pituitin," and affect vomiting, pus, ulcers, the uterus, bedbugs, migraines, etc.

Tagetes

1. Used for colds, gastro-intestinal disorders, and fever.

2. Has antibacterial and diuretic activity.

3. Handling marigold can cause dermatitis.

Its greatest utility was in curing rash and skin infections. It could accomplish all this through being 'hot' and 'dry' in the fourth degree, thus counteracting 'cold' diseases, according to Hernández (1959, 2:324). Before the Spanish Conquest, *yyauhtli* was associated with the powerful water god Tlaloc (Ortiz de Montellano 1990).

Hernández (1959, 2:218–21) described seven principal varieties of other *Tagetes*. One was *cempoalxóchitl*, considered 'hot' and 'dry' in the third degree. The juice of the leaves helped the 'cold' stomach, provoked menstruation, urine, and sweat, alleviated "intermittent chills" if applied before the attack, expelled flatus, cured weakness coming from 'coldness' of the liver, opened the veins, alleviated hydropsy, provoked vomiting, and cured fever.

Tagetes species were given the common name of *anís* by the Spaniards in northwestern New Spain because the root smelled like anise. The Ópata chewed the root of *anís* to aid digestion and ease discomfort caused by overeating, according to Nentuig ([1764] 1977). Pfefferkorn ([1794–95] 1989) said that a dangerous inflammatory fever could be cured by drinking a decoction of the plant instead of by bloodletting. Esteyneffer ([1719] 1978) recommended the decoction for suppressed urine, as a fomentation for hemorrhoids, and in an enema for detained menses, for *mal de madre* if the woman was near death (otherwise as a drink), and if the woman was in labor for more than twenty-four hours. Longinos in his 1792 journey (Engstrand 1981) found a species of *Tagetes* used as a strong stimulant in Baja California.

Modern Use. The Tepehuan use *Tagetes jaliscana* (common name not given) to relieve stomach disorders. The Mountain Pima make a drink of *T. elongata* or *T. micrantha* for stomach indisposition and use *T. lucida* in a drink with *mezcal* for high fever. The Tarahumara also make extensive use of *T. lucida* and *T. micrantha*, employing teas made from the leaves or the whole plant for gastrointestinal disorders, headache, colds, chills, pneumonia, and fever. The Warijio make a tea of the dried herbage of *T. filifolia* to relieve minor indispositions and *T. lucida* for headache and stomachache.

Medicinal *yerba anís* is sold in herb stores and certain

Arizona supermarkets. Mexicans make a tea from the leaves and flowers. For a cold Mexican Americans take a tea brewed from the leaves every four hours. If the cold is accompanied by a runny nose and fever, rum is added. They also plant *Tagetes* in their gardens to protect roses from attack by aphids.

Phytochemistry. According to NAPRALERT, numerous *Tagetes* species have been studied, especially *Tagetes erecta*, in countries including Argentina, Denmark, Egypt, France, Germany, India, Thailand, and the former Soviet Union. The essential oil has been found active against *Pseudomonas aeruginosa, Staphylococcus aureus, Bacillus cereus,* and various fungi. *Tagetes minuta* was successfully tested as an insecticide; it contains pheromones, which protect plants from destruction by insects. Many compounds have been found in the leaves and flowers of *Tagetes* species, including linalool, luteolin, coumarins, alpha- and beta-pinenes, stigmasterol, and sitosterol. *Tagetes erecta* and *T. lucida* contain quercetins and kaempferol (Ortiz de Montellano 1990:253). Both the leaves of *T. erecta* and the leaves and flowers of *T. lucida* have been found to inhibit *Streptococcus pyogenes* (Cáceres et al. 1991). In another study, *T. lucida* was found to be effective against *Salmonella typhi* and *Pseudomonas aeruginosa* (Cáceres et al. 1993).

■ *Tecoma* (Bignoniaceae)
TRUMPET BUSH

Tecoma stans (L) H.B.K.
  nextamalxóchitl oapanense, Aztec; *kusi urákame,* Tarahumara; *palo de arco,* Baja California Sur; *tronadora,* Baja California Norte, Spanish; trumpet bush, English

Trumpet bush is a small, delicate tree with funnel or trumpet-shaped, fragrant yellow flowers and narrow, beanlike seedpods. It is a popular ornamental. The genus name is from the Nahuatl *tecomaxochitl* (*tecomatl,* clay pot; *xochitl,* flower).

Historic Use. Hernández (1959, 3:59) recommended *nextamalxóchitl* for fever, exanthems, and other 'hot' conditions because of its 'cold' nature.

FIGURE 29 *Tecoma stans* var.
angustata. a. Fruit.

Tecoma

1. Used for stomachache
and diabetes, kidney, and
liver disease.

2. Has antiinflammatory,
hypoglycemic, and anti-
biotic action.

3. No reports of toxicity.

Modern Use. The Tarahumara use the flowers in a tea for
colds and in an ointment for heart ailments. In Baja California Sur, a decoction made of the flower, branches, and
bark is taken to control blood pressure, diabetes, and stomachache. In Baja California Norte and much of the rest of
Mexico (Linares, Bye, and Flores 1984) the plant is known
as a remedy for diabetes, liver disease, and gastritis. Mexican
Americans can obtain *tronadora* in certain supermarkets.

Phytochemistry. Phytochemical research conducted on
Tecoma stans (see Winkelman 1986:123, 1989:264) shows
that the plant contains coumarin, tecomanine, and tecosta-
nine, with hypoglycemic actions; it was found to be one of
the best antidiabetic plants in a sample of twenty-one Mexican plants surveyed (Pérez et al. 1984). It contains chloro-
genic acid, coumaric acid, oleanolic acid, and beta-sito-

sterol, which affect the liver. It also contains lapachol and phloretic acid, which are bactericidal.

■ *Turnera* (Turneraceae)

DAMIANA

Turnera diffusa Willd. ex Schult.
tabuquit, Ópata; *damina,* Pima Bajo, Yaqui, Baja California; *damiana, yerba del pastor, hierba de la pastora, hierba del venado, pastorcita,* Spanish; damiana, English

Turnera ulmifolia L.
saráame, Tarahumara

Damiana is a shrub with small, whitish leaves and yellow flowers.

Historic Use. In northwestern New Spain, Nentuig ([1764] 1977:63) said that a priest of known honesty and much experience told him that drinking a decoction of damiana made women fertile whom he had seen previously to be sterile and infecund. Pablos, reporting to his superior in 1784, said that *yerba del pastor* should be fried in wax, then applied to the groin to treat fever and urinary disease.

Modern Use. The Yaqui make a tea from the plant of *T. diffusa* for stomachache. The Tarahumara drink a tea of the leaves of *T. ulmifolia* for diarrhea. Damiana is best known for treating sexual problems. Pima Bajo women who cannot conceive use the tea of *T. diffusa* to promote conception. In Baja California Norte *T. diffusa* is used for impotence, frigidity, sterility, and sexual exhaustion, as well as for diabetes. In Baja California Sur the decoction of the branches and leaves is employed to fortify the uterus, for cold and cough, and also for scorpion stings. Mexican American women say it will correct *frío en la matriz* ('cold' uterus, infertility) and inflammation. Proprietors of herbal markets in Mexico recommend mixing it with *gobernadora* (*Larrea,* creosote bush) and *mariola* (*Parthenium* sp.). A tea made from this mixture is to be taken before breakfast and at night to aid conception. This is one of the most sought-after herbs in the Tucson *botanica.* It also sells well in certain supermarkets.

Turnera

1. Believed to aid conception and treat diabetes.

2. Has effective hypoglycemic action.

3. Contains cyanogenic glycoside and arbutin but is approved for food use.

Phytochemistry. The essential oil of *T. diffusa* contains cineol, alpha-and beta-pinenes, and beta-sitosterol (Duke 1985:492). No compound has been identified that explains use of *Turnera* species as aphrodisiacs (Tyler 1993:108). *Turnera diffusa* was one of the most effective antidiabetic hypoglucemics in a sample of twenty-one Mexican plants (Pérez et al. 1984). It contains cyanogenic glycoside and arbutin but is nevertheless approved for food use. An extract of the branches was found to have 1 + effectiveness against *Staphylococcus aureus* and 2 + effectiveness against *Bacillus subtilis* (Encarnación and Keer 1991).

■ *Valeriana* (Valerianaceae)
VALERIAN, GARDEN HELIOTROPE

Valeriana ceratophylla H.B.K.
valeriana, Spanish

Valeriana edulis Nutt.
chinonua, Tarahumara

Valeriana officinalis L.
valeriana, Spanish; valerian, English

Valeriana spp.
cuitlacamotli, Aztec; *valeriana,* Baja California Norte; *valeriana, hierba de los gatos, raiz de gato,* Spanish; valerian, English

Valerians are large herbaceous plants with single stems but highly divided leaves. The name probably comes from the Latin *valere* (to be healthy), referring to medicinal properties. The Aztec name denotes its strong odor (*cuitla-*) and the tuberous root shape (*camotli,* as in *camote,* sweet potato).

Historic Use. Valerian was labeled with the designation *officinalis,* showing that it was long employed in Old World medicine. Gerard ([1633] 1975:1078) noted many uses for it, but not as a sedative or soporific. The Aztecs mashed the root of a *Valeriana* species, using it to induce urination and cure kidney diseases. Hernández (1959, 2:224) categorized it as mucilaginous with some 'heat'. In northwestern New

Spain, Esteyneffer ([1719] 1978) gave the name *valeriana* as a synonym for *yerba del manso*, although he recommended the plant for purposes consistent with those of *yerba del manso (Anemopsis)*, not valerian—including dental hemorrhage, head wounds, and genital ulcers. True *Valeriana* seems not to have been used in the eighteenth century.

Modern Use. The Tarahumara make a tea from the root of *V. edulis* for medicine. The Mayo make a tea from the leaves of a valerian for "nerves" and for stomachache. In Baja California Norte, a valerian is recommended for hysteria, stress, insomnia, and nervous conditions, and as an antispasmodic and sedative. Mexican Americans also drink valerian tea to calm and as a soporific, purchasing it in grocery and herb stores. It is becoming popular with Euro-Americans.

Phytochemistry. *Valeriana* contains iridoid compounds as well as several alkaloids (Duke 1985:504). The root and rhizome produce an essential oil that decomposes into valeric acid, a soporific and an antispasmodic, and methyl ketone, a mild anesthetic (Schauenberg and Paris 1977:258). However, one researcher (Tyler 1993:316) has stated, "At the present time, the identity of the principles responsible for valerian's therapeutic effects remains unknown." Because of its strong odor and taste it is sometimes taken in capsules. It has been approved, Generally Recognized as Safe, by the FDA.

Valeriana

1. Used to calm and as a soporific.

2. Has antispasmodic, soporific, and anesthetic action.

3. No reports of toxicity.

■ *Vallesia* (Apocynaceae)

VALLESIA

Vallesia glabra (Cav.) Link
huitatobe, Cáhita; *tudog us, palo verde*, Pima Bajo; *duduwasa*, Tarahumara; *sitavaro*, Warijio; *citabaro*, Mayo; *tonóopa, tinóopa, huevito*, Seri; *otatave, huitatave, cacaragua*, Spanish

Vallesias are shrubs containing a white latex. The stone fruits are small, with white or pinkish flesh.

Modern Use. The Pima Bajo squeeze the fruit to obtain a liquid that is applied to inflamed eyelids. For pinkeye and

Vallesia

1. Used to treat eye conditions, rheumatism, and skin conditions.

2. May have bacteriocidal action.

3. No reports of toxicity, but contains the sedative reserpine.

other eye diseases, the Warijio insert the juicy pulp in the eye; they burn the foliage and branches to ashes and rub the ashes on itches and on measles. The Mayo roast vallesia on a grill or in hot ash, being careful not to burn it, and then place it with a rag on the body part painful with rheumatism. For sores inside the body, they cook the most tender parts of the shoots. The Tarahumara use the fruit for medicine. The Seri burn the leaves until blackened, grind them into a powder, and rub this powder on the body part inflicted with rash or severe itching such as from measles. This plant has not remained in the plant pharmacopoeia of contemporary Mexicans and Mexican Americans, perhaps because they have substituted more satisfactory medicines for problems formerly treated with vallesia.

Phytochemistry. *Vallesia glabra* was not reported by NAPRALERT as having been studied. However, *Vallesia dichotoma* was found to be used as a bactericide and for inflamed eyes in Peru; in *Vallesia antillana* of Cuba, indole alkaloids were found to be present in the stem bark, leaf, branches, and root. The sedative reserpine has been noted as a toxin in *Vallesia* species (Duke 1985:568).

■ *Zea mays* (Poaceae)

CORN, MAIZE

Zea mays L.
tlaolli, maíz, Aztec; *maíz,* Mountain Pima, Tarahumara, Mayo; *maíz, elote* (ear of tender corn), *barba de elote* (corn silk), Spanish; corn, maize, English

Maize is a large plant with grains borne on a side branch, the cob. It is one of the world's most widely used food crops. This grain is used by nearly every agricultural Native American group as food, and by many as medicine. It is sacred to many.

Historic Use. Hernández (1959, 2:288–92) wrote of the great variety of dishes to be made from the different kinds of *tlaolli,* called *maíz* by Haitians. *Xocoatolli* was made by mixing a pound of fermented black corn meal with two pounds of white corn meal, adding salt and chili. When eaten in the morning, it was thought to clean and purge the body and provoke urine. *Tlatonilatolli*—made by mixing a little corn, chili pepper, and some *epazote* and cooking the mixture well—was said to excite sexual appetite, provoke urine and menstruation, and heat and strengthen the body. Corn, he said, was temperate, somewhat inclined to be 'hot' and 'moist'.

Hernández also described another food, made by "the *chichimecas*, the fierce, barbarous and unconquered who live not far to the north and roam the mountains and fields covering only parts of their body with animal skins," subsisting mainly on hunting. This dish was meat covered with corn masa, baked in a pit lined with hot stones. The result was so delicious that Hernández began to introduce the practice to the Spanish residents of Mexico. (Tamales, the modern equivalent, are enjoyed throughout the United States and Mexico today).

Since Hernández, many other medicinal uses for corn have been noted. In northwestern New Spain, the Ópata were reported to make a thick poultice of corn meal that they applied cold to treat the throat swelling of *garrotillo.* Among many uses for *maíz*, Esteyneffer ([1719] 1978:846) recommended consuming *atole*, thin corn gruel, for diarrhea. He applied the toasted and powdered kernels to bruises.

Modern Use. The Tarahumara use the styles (silk) of corn in tea for urinary complaints. The Mountain Pima add filaments of ocotillo to corn silk for a tea to treat pneumonia. The Mayo take a gruel made from corn meal for diarrhea. Mexican Americans use corn silk, *barba de elote,* in an tea that they drink for urinary complaints. To effect a cure, the tea must be taken without the customary honey or sugar for one month. They grind the kernels into flour to make an *atole* for diarrhea. An infant's diarrhea is treated by making the atole with two tablespoons of corn starch and adding cinnamon. *Barba de elote* is one of the most commonly sold plant remedies at a Tucson *botanica* and certain supermarkets.

Phytochemistry. The use of *Zea mays* as a medicinal has spread all over the world. Twenty-six pages of phytochemical research are summarized in the NAPRALERT report.

Zea mays

1. Corn silk is used to treat urinary complaints, corn flour for diarrhea.

2. Relaxes the ureter and is diuretic.

3. Contains various toxins, but there are no reports of toxicity.

Compounds that explain medicinal properties have been identified in the pollen, seed, oil, and styles. *Zea mays* has been found to have antispasmodic activity, which might help with an infant's diarrhea. It is also a smooth muscle stimulant yet has relaxant activity of the ureter as well as diuretic activity, which might explain its use in urinary complaints. The silk contains cineol, alpha-terpineol, allantoin, and tannins, as well as alkaloids, carotenes, and coumarins. The seed oil is high in steroids and contains cycloasdol, daucosterol, eugenol, and limonene.

■ *Zexmenia* (Asteraceae)
ZEXMENIA

Zexmenia aurea Benth. & Hook
atepocapatli, medicina de renacuahos, Aztec; *peonía,* Mountain Pima

Zexmenia podocephala A. Gray
gogóši viítai (dog excrement), Tepehuan; *reyóchari,* Tarahumara; *pioniya, pionilla,* Mountain Pima, Warijio

Zexmenia seemanni A. Gray
wachomo', Warijio

Zexmenia spp.
la peonía, peonilla, Ópata, Mayo, Spanish

Zexmenia is a perennial herbaceous plant with yellow flowers and a tuberous root. It is found in mountains at altitudes above 4,000 feet.

Zexmenia

1. Used for diarrhea and other gastrointestinal problems.

2. No reports of biological activity.

3. Contains sesquiterpene lactones and saponins, which are toxic.

Historic Use. The Aztecs mashed the root of *atepocapatli* (*atepoca-*, tadpole, for the shape of the rhizome; *patli-*, medicine) to heal problems of the stomach and kidney, as well as sterility caused by 'cold'. Hernández (1959, 2:48–49) reported that some Indian doctors whom he consulted said that the decoction also extinguished fever, which he could not understand unless there might be hidden some 'cold' parts in the plant, since the roots, he declared, were 'hot' and 'dry' in the third degree.

In northwestern New Spain, Sonorans made a drink by boiling the bitter fresh root and skin of a *peonilla,* or drying it and taking it in pulverized form for a disordered stomach, or chewing the root and swallowing the saliva (Pfeffer-korn [1794–95] 1989:62). Nuñez (1777), describing the plant as different from the peony of Spain, recommended adding sugar to the brew for medicine.

Modern Use. For stomach disorders, the Mountain Pima make a drink of *Z. aurea*. They collect the roots of *Z. podo-cephala* to sell in Sonora for preparing a drink to treat se-vere stomach cramps. The Tepehuan crush the roots of *Z. podocephala* and add this to warm water for a purgative drink, although they claim the root is poisonous even in small amounts. The Tarahumara use the root tea of this species for gastrointestinal problems and rheumatism. The Mayo use the root tuber of a *Zexmenia* species for *empacho*. The Warijio decoct the tuberous roots of *Z. podocephala* for stomach ailments and the leaves of *Z. seemanni* to facilitate menses or induce labor in childbirth. Mexicans buy the root of *peonía* to make a decoction for diarrhea, and it was also used by Mexican Americans, who now prefer over-the-counter remedies.

Phytochemistry. According to NAPRALERT, other species of *Zexmenia* have been studied, but not the species known to be used in the American and Mexican West. The aerial parts contain lupeol, valnerol, and numerous sesquiter-penes, diterpenes, and triterpenes. Sesquiterpenes, sulfur compounds, and diterpenes have been found in the root. *Zexmenia* species contain sesquiterpene lactones and sapo-nins, which are toxic. No information on biological activity is given in NAPRALERT.

■ *Zornia* (Fabaceae)
SNAKEWEED

Zornia diphylla (L.) Pers.
koi vasogadi (vibora yerba), Tepehuan; *yerba de la víbora*, Spanish

Zornia reticulata Sm.
yerba de la víbora, Mountain Pima, Tarahumara

Zornia spp.
yerba de la víbora

Zornia species are small herbaceous plants with stiff leaves bearing yellow gland dots. The yellow flowers are born on spikes.

Zornia

1. Used for colds and stomachache.

2. Other species show antibiotic activity.

3. No reports of toxicity.

Historic Use. Pablos, stationed at Rio Chico in 1784, was the first of the sources used in this study to tell of a *yerba de la víbora* that was a sudorific for fever and chills. Other writers of medical botany in the eighteenth century described various *yerbas de la calentura* (fever herbs), but it is not possible to know which plants they meant.

Modern Use. The Mountain Pima boil the leaves of *Z. reticulata* to make a drink for colds and stomachache. The Tepehuan use a tea made from the plant of *Z. diphylla* to relieve pains caused by the grippe. The Tarahumara use *Z. reticulata* to make a tea to treat chills and fever; the plant is also sold for this purpose in Mexico City. Mexican Americans in Las Cruces, New Mexico, brew the leaves of *yerba de la víbora,* adding honey with lemon, to treat a cold or a stomachache. It is said to cause healthful sweating if there is a fever; the patient takes the tea every four hours and should be protected from exposure to air. *Yerba de la víbora* is sold in Mexican grocery stores and in certain southwestern supermarkets.

Phytochemistry. Little is known about these *Zornia* species. According to NAPRALERT, studies of other species in Brazil, Costa Rica, India, and Kenya have showed antibacterial activity against *Bacillus subtilis, Mycobacterium smegmatis,* and *Staphylococcus aureus.* In one study, *Z. diphylla* had equivocal activity against *Helminthosporium turcicum* and demonstrated the presence of coumarin.

APPENDIX A
SAFETY OF MEDICINAL PLANTS

"Wrong person, wrong dose, wrong combination." Any medicinal may be unsafe if taken by a sensitive person, if the dose is incorrect, or if other medicines that may potentiate or act synergistically are also taken. These facts must be emphasized: just as any biomedical pharmaceutical can under certain circumstances be dangerous, there can be untoward reactions to any plant medicine. One cup of a tea might help: more may make one sick. Parts of one-fourth of the 100 plant genera used for medicines in the American and Mexican West are known to be toxic when taken internally, as follows:

Argemone	*Mentha* (oils)
Aristolochia	*Mentha pulegium*
Asclepias	*Nicotiana*
Caesalpinia	*Perezia*
Chenopodium (oil)	*Phoradendron, Struthanthus, Loranthus*
Datura	*Pinus* (gum)
Euphorbia	*Plumeria*
Heterotheca	*Psacalium*
Jatropha	*Rhynchosia*
Juniperus	*Ricinus* (beans)
Karwinskia	*Sambucus* (stems)
Krameria	*Senecio*
Lantana	*Solanum*
Larrea	

Nearly one-fifth of the plants have been studied insufficiently to evaluate their safety. In some cases, other species in the genera have been studied, but not the particular species that are used medicinally in the American and Mexican West.

Agastache	*Ligusticum*
Anemopsis	*Mammillaria*

Arracacia

Buddleia

Gutierrezia

Haplopappus

Ibervillea

Jacquinia

Kohleria

Mascagnia

Ruellia

Rumex

Sphaeralcea

Vallesia

Zexmenia

One-fifth of the genera include plants that may cause dermatitis or other allergic reactions:

Acacia

Achillea

Agave

Ambrosia

Artemisia

Asclepias

Cannabis

Capsicum

Cassia

Euphorbia

Eucalyptus

Jatropha

Juniperus

Matricaria

Mentha

Prosopis

Rosa

Rosmarinus

Ruta

Tagetes

Pregnant women and nursing mothers should avoid the toxic genera listed above. Moreover, pregnant women should particularly avoid the following plants, which are especially notorious as abortifacients:

Achillea

Aloe

Eryngium

Guaiacum

Gutierrezia

Juniperus

Karwinskia

Perezia

Phoradendron

Physalis

Rosmarinus

Ruta

APPENDIX B
PHARMACOLOGICALLY ACTIVE
PHYTOCHEMICALS

acacetin: antiinflammatory

allantoin: antiinflammatory, suppurative

allicin: antitumor, bactericide, fungicide, hypoglycemic, hypocholesterolemic,
   insecticide, larvicide

anisaldehyde: insecticide

anthraquinones: laxatives

apigenin: antispasmodic

arborine: abortifacient

astragalian: immunostimulant

aucubin: antibacterial

azulene: antiinflammatory, antipyretic

benzoic acid: antifungal, antipyretic, antiseptic, expectorant

berberine: amoebicide, bactericide, anticonvulsant, antidiarrheal, antiinflammatory,
   astringent, candidicide, carminative, collyrium, febrifuge, fungicide, hemostatic,
   herbicide, immunostimulant, sedative, trypanosomicide, uterotonic

α-bisabolol: antiinflammatory, antiulcer, bactericide, fungicide

chamazulene: anodyne, antiinflammatory, antiphlogistic, antispasmodic, germicide

chrysophanic acid: laxative

cineol: expectorant for respiratory conditions (bronchitis, laryngitis, pharyngitis,
   rhinitis), insecticide

coumarin: antitumor, antiinflammatory, hypoglycemic

cresol: expectorant

cycloasdol: antiinflammatory

damsin: antitumor

daucosterol: antitumor

diosgenin: antiinflammatory and estrogenic

dopamine: adrenergic, antihypotensive, antilactagogue, antiparkinsonian

emodin: antimicrobial, purgative

eugenol: analgesic, antiseptic, anesthetic, fungicide, larvicide

furanoquinoline: abortifacient

gitogenin: antiinflammatory
heliamine: antitumor
hispidulin: antitumor
kaempferol: antiinflammatory, diuretic, natriuretic
lapachol: antitumor, bactericide, fungicide
linalool: anticonvulsant, antimicrobial, spasmolytic
lupeol: antitumor
luteolin: antiinflammatory, antispasmodic, antitussive
methyleugenol: sedative
methyl salicylate: analgesic, antiinflammatory, antipyretic, antirheumatic
parthenolide: antitumor
protopine: analgesic, convulsant, hypotensive, sedative
quercetin: antiinflammatory, antispasmodic
rumicin: antiparasitic
rutin: antiatherogenic, antiedemic, antiinflammatory, antithrobogenic, hypotensive,
    spasmolytic, vasopressor
salicylic acid, salicin: antipyretic, analgesic, antirheumatic
sanguinarine: anesthetic, antiseptic, antitumor, fungicide, saliva producing
saponin glycosides: antibiotic and fungistatic activity
β-sitosterol: antihypercholesterolemic, antiprostatic, antiprostatadenomic, estrogenic
solanine: antiasthmatic, antibronchitic, antiepileptic
solasodine: antiandrogenic, antiinflammatory, antispermatogenic
tannins: antidiarrhetic, bactericide, viricide
terpineol: antiallergenic, antiasthmatic, antitussive, bacteriostatic, expectorant
thujone: uterine stimulant, abortifacient
thymol: bactericide, fungicide, larvicide, tracheal relaxant, vermicide
tigogenin glycoside: antitumor
valnerol: sedative

Source: Duke 1985:569–79, table 4.

NOTES

INTRODUCTION

1. Edward Spicer (1980:xiii) states that "the most vital element in the modern discipline of ethnohistory . . . consists in the interpretation of documented events of the past by means of the knowledge of situations which anthropologists have gained through direct study of living societies . . . [so that] anthropologists may assume uniformities running through the whole experience of humans—past, present, and future."

2. In one survey (Eisenberg et al. 1993), one-third of the people in the sample used alternative medicine, generally in addition to conventional therapy and commonly without telling their doctors.

3. In homeopathic remedies, plant parts are diluted to an infinitesimal amount, and an entirely different theory of phytotherapy underlies the homeopathic system. For example, *Ruta graveolens*—used in the American and Mexican West to treat earache or provoke abortion—is prescribed by homeopathic physicians to treat lower back pain. Homeopathy employs many of the plants that are discussed in this book, including *Acacia, Achillea, Agave, Aloe, Anemopsis, Argemone, Aristolochia, Artemisia, Asclepias,* cactus species, *Cannabis, Cassia,* chamomile, *Chenopodium, Datura, Ephedra, Eryngium, Euphorbia, Gnaphalium, Guaiacum, Haematoxylon, Jatropha, Juniperus, Krameria, Lippia, Mentha, Paeonia (Zexmenia), Plantago, Plumeria, Quercus, Rosa, Rosmarinus, Rumex, Ruta, Salix, Salvia, Sambucus, Senecio, Solanum, Turnera, Valerian, Verbascum, Zea.*

4. For this look at change in medical culture I return to an older definition of acculturation (Redfield, Linton, and Herskovits 1936, cited in Beals 1953): "Acculturation comprehends those phenomena which result when groups of individuals having different cultures come into continuous first-hand contact, with subsequent changes in the original cultural patterns of either or both groups." More recently acculturation has been defined as occurring only in the direction of the dominant group.

5. I have conducted interviews since beginning research in 1960; the most recent as of this writing was September 15, 1995.

6. These include morphine, codeine, quinine, atropine, digitalis, vinblastine, vincristine, and aspirin (Abelson 1990:513).

7. See especially Trotter 1981a, 1981b; Trotter and Chavira 1981. Also see Saunders 1954; Foster 1953; Clark 1959; Rubel 1960; Madsen 1964; Curtin 1965; Kelly 1965; Madsen 1968; López Austin 1970, 1980; Kay 1972; Ford 1975; Achor 1978; Meredith 1982; Anzures y Bolaños 1983; Bastien 1987; Orellano 1987; and Ortiz de Montellano 1990. See bibliographic essay for details.

8. See Krochmal and Krochmal 1973; Moerman 1991; Foster and Duke 1990.

9. See Lozoya, Velázquez, and Flores 1988; Lozoya 1990.

10. See Velimirovic 1978; Lozoya, Velázquez, and Flores 1988.

11. Although some plants may contain toxic compounds, few are actually fatal when ingested in small or moderate amounts according to Kingsbury (1964).

1: ETHNOHISTORY

1. Some theorists think medicinal plant knowledge expanded with the development of agriculture, others that it was discovered secondarily to the search for food or in some cases that using a plant to heal led to using it as food (see Etkin 1994).

2. See Greenberg, Turner, and Zegura 1986; Greenberg and Ruhlen 1992.

3. In support of this premise, many medicinal plants now used in Alaska are also employed in Russian folk medicine, including plants used everywhere in herbal medicine as well as plants that are endemic to their similar climates. See Kourennoff 1970; Hutchens 1973; Fortuine 1988; Graham 1985; Kari 1991.

4. Pliny 1938, 7:201. Research in zoopharmacognosy (recognition and utilization of medicinal plants by animals) finds that elephants, monkeys, bison, pigs, civets, jackals, tigers, bears, wild dogs, rhinoceros, mole rats, and desert gerbils appear to ingest plants for medicines (Janzen in Johns 1990:253).

5. See Leroi-Gourhan 1975; Solecki 1975; Lietava 1992.

6. See Reinhard, Ambler, and McGuffie 1985; Reinhard, Hamilton, and Helvy 1991; Trigg et al. 1994.

7. See Dean 1993:112. It is interesting, however, that Galen, who wrote in the second century A.D., recommended slitting willow bark for medicine while the tree is in flower, to collect "juice" to treat eyes (Gerard [1633] 1975:1392).

8. "This interaction encompassed not only the exchange of goods but also of ideas, accounting for shared beliefs, symbols, and architectural forms in the two regions" (McGuire 1980:32).

9. Alaska shares medicinal use of genera commonly employed in the American and Mexican West such as *Achillea, Arctostaphylos, Artemisia, Equisetum, Juniperus, Pinus, Plantago, Populus, Rosa, Rumex, Salix, Sambucus, Senecio,* and *Valerian,* plants that are used as medicines everywhere that they can flourish.

10. See Foster 1953, 1994; MacFarlane 1962; Anderson 1987; Kay and Yoder 1987.

11. "But the reason why more herbs are not familiar is because experience of them is confined to illiterate country-folk, who form the only class living among them; moreover nobody cares to look for them when crowds of medical men are to be met everywhere" (Pliny 1938, 7:147).

12. *The Travels of Marco Polo* (1958), a translation of a fourteenth-century manuscript in the Bibliothèque Nationale, Paris, has frequent references throughout to traders of "spiceries and drugs" as well as other curative devices.

13. Morison (1963:78 n.3) notes that Marco Polo wrote of a tree that he saw in Malaysia, lignum aloe *(Aquilaria agallocha),* which is unrelated to either *Aloe* or *Agave.*

14. The Spaniards soon introduced aloes into the West Indies, where they are extensively cultivated. Monardes (1577) reported that don Francisco de Mendoza attempted to transplant cloves, pepper, and ginger from the oriental Indies to New Spain, succeeding only with ginger.

15. This statement is widely quoted but has proved elusive; many of us have looked but have not actually found it in a letter.

16. Francisco López de Villalobos, author of *El sumario de la medicina con un tratado de las pestiferas bubas,* was a physician from a family of *conversos.* Safely em-

ployed as a physician to Ferdinand and then to Charles V, he nevertheless had difficulties caused by his former religious identity.

17. It was in a state of neglect when I visited Cuernavaca at the time of the second Congreso Internacional de Medicina Tradicional y Folklórica.

18. Nevertheless, the practice of invoking a supernatural is cross-cultural: the Catholic friars prayed to a special saint for each disease.

19. Clavijero (1787:426n. c) writes of Hernández, "His work . . . consisted of twenty-four books of history, and eleven volumes of excellent figures of plants and animals; but the king thinking it too voluminous, gave orders to his physician Nardo Antonio Ricchi . . . to abridge it. This abridgement was published in Spanish by Francisco Ximenes, a Dominican, in 1615, and afterwards in Latin, at Rome, in 1651, by the Lincean academicians, with notes and learned dissertations. . . . The manuscripts of Hernandez were preserved in the library of the Escurial, from which Nuremberg extracted . . . a great part of what he has written in his Natural History." Unfortunately, the manuscripts that had been thought to be preserved in the Escorial were in fact destroyed by a disastrous fire there June 7, 1671 (Martínez 1969).

20. Ximénez, in his foreword to his 1615 publication, takes these doctors to task for appropriating and corrupting the work of Hernández. In fact, Barrios does acknowledge Hernández.

21. See also Comas 1954, 1964, 1968, 1971.

22. The official name for this plant is now given as *Dorstenia drakena* L., entered by Sessé and Moziño ([1787] 1887), and also appears as *Dorstenia contrajerva* (Martínez 1979). It may be purchased today in Tucson.

23. His work consisted mainly of translations from the works of the Dutch Doedoens by a Dr. Priest.

24. Sir Hans Sloane, who studied at the famous Chelsea Physic Garden during his early training as a physician, also has its refounding in 1722 to his credit. His statue can be seen in its center.

25. I have made this as well as other translations from Spanish or German.

26. For example, a popular German herbal (Schönfelder and Schönfelder 1981) describes only 6 New World plants out of 442 medicinal plants.

27. My summaries do not include Californian Indians or Colorado River groups. I have conducted no studies with these peoples, but ethnobotanists such as Almstedt (1977) for the Digueño, Heiser and Elsasser (1980) for many Californians, Timbrook (1987) for the Chumash, and Zigmond (1981) for the Kawaiisu describe similar plant uses, which occurred especially after contact with Spanish missionaries.

28. His "household remedies" included animal excreta and other materials that are unusual from a twentieth-century perspective.

29. For various interpretations of archaeological data concerning the prehistory of this region see McGuire and Schiffer 1982; Mathien and McGuire 1986; and Woosley and Ravesloot 1992.

30. Ethnobotanists of the Tarahumara include Bennett and Zingg (1935), Pennington (1963), and Bye (1976, 1979, 1986), all of whom have deposited voucher specimens.

31. Information on medicinal plant use by the Mayo comes from data collected in the *municipio* of Masiaca (Kay and O'Connor 1988) and in Las Animas, near Huatabampo (Cozarit 1985). Thomas Van Devender and David Yetman are conducting a new study.

32. For Yaqui medicinal plant use see Wagner 1936; Shutler 1977; Williams 1983; Felger, personal communication 1993.

33. Clavijero (1787:426–27) writes, "Among other arts exercised by the Mexicans,

that of medicine has been entirely overlooked by the Spanish historians" but not by the scientists, since from "the natural history of Mexico, written by Dr. Hernandez . . . a system of practical medicine may be formed for that kingdom; as has in part been done by Dr. Farfan in his Book of Cures, by Gregorio Lopez, and other eminent physicians."

34. For Baja California Norte, see Winkelman 1986; for Baja California Sur, see Encarnación and Agundez 1986; Encarnación, Fort, and Luis 1987; Encarnación and Keer 1991.

35. See Hammond and Rey 1953:104. Most of the pharmaceuticals listed were mineral remedies such as lead, mercury, and sandalwood ointments; other ointments with names like apostolorum, egyptian, incarnative, and calmative; rock sulfur; white coral; and alum. Medicines made from plant materials included coriander, pomegranate, rose, and violet syrups; borage and fennel water; chamomile, rose, myrtle, quince, and rose oils; sweet clover, granulated aloes, centaury, marshmallow, citrus, and apple paste—medicines commonly used in Europe and America until this century.

36. I found almost all the plants I have discussed in this book, and more, on visits to a Tucson *botanica* and to stores in a supermarket chain in October 1995.

2: PLANTS, THEIR NAMES, AND THEIR ACTIONS

1. Analyzing and drawing conclusions about plant use as in this book requires that one carefully examine certain assumptions about the names: (1) common names of plants are stable, that is, they are not attached idiosyncratically to various plants but speak to cultural practice; (2) the plants have been described accurately in the sources; (3) the names given to illness conditions that the ethnographer records accurately reflects the informant's beliefs; and (4) all members of a culture group would agree with the medicinal plant names and uses given by the key informants, those persons with special knowledge.

2. It is interesting that when there are differences, in most instances they reflect using different species of the same genus. With the work of Bye (1976) and Laferrière (1991), other correspondences of medicinal plants have been found.

3. See Gates 1939; Ortiz de Montellano 1976.

4. Hernández (1959, 2:7–8) wrote that the plant is called *acocoxóchitl.* It is illustrated by a drawing that appears to be the plant we now know as *Dahlia coccinea* Cav.

5. The illustrations in the *Badianus* of Martin de la Cruz, though elegant and profuse, are too stylized to make accurate identifications with certainty (see Emmart 1940; Gates 1939).

6. See bibliographic essay for details.

7. See bibliographic essay for details.

8. A few plants, such as *jojoba* and *zapote,* are important enough to be included even if they were reported by fewer than four ethnic groups.

9. This section derives from Schauenberg and Paris 1977; Duke 1986; Tyler, Brady, and Robbers 1981; Fuller and McClintock 1986.

10. Interested readers are referred to Schauenberg and Paris 1977; Font Quer 1979; Tyler, Brady, and Robbers 1981; Lewis and Elvin-Lewis 1982; Tyler 1993. The *Journal of Ethnopharmacology* and the journal of the American Botanical Council, *HerbalGram,* are recommended as well as the data bank of the University of Illinois, NAPRALERT.

3: ILLNESSES TREATED WITH PLANTS

1. For more particulars on disease names, see Kay 1976, 1979, 1993.

2. See López Austin 1980 for analysis based on religious symbolism. See also Ortiz de Montellano 1990; cf. Foster 1994.

3. For sources of information on each culture group's use of medicinal plants in this data bank, see bibliographic essay. My data for Aztec plants come from the *Obras completas* ([1571–1576] 1959–1976) of Hernández, who detailed the qualities of most medicinal plants that he studied in sixteenth-century Mexico as to the degree of heat, cold, moistness, and dryness, as well as additional properties such as glutinous, acrimony, etc. I have used the work of Hernández extensively because his writings, errors and all, appear to have greatly influenced the contemporary ethnic medicine of Mexican Americans. References to the ancients such as Dioscorides and European Renaissance writers—principally Gerard in *The Herbal* ([1633] 1975)—also describe the humoral complexion of each plant.

4. From *Medicinal Herbs of Early Days of Mission San Antonio de Padua,* a photocopy of studies of Father Doroteo Ambris (d. 1882) in southern Monterey County, California, edited by Father Zephyrin Engelhardt.

5. Bye and Linares (personal communication) report that as recently as 1990, *guareke* was unknown; yet by December 1994 Mexico City markets were selling these giant tubers at prices up to $20 (U.S.). In Arizona, a packet of thirty capsules cost the customer $7.50 in October 1994.

4: HEALING THE ILLNESSES OF WOMEN AND CHILDREN

1. This chapter includes information from my two data assemblages; not only Southwestern Medicinal Plants, which was generally used in this book, but also Ethnoagents in Women's Health Care, which summarizes women's remedies of 170 ethnic groups, including those of the American and Mexican West (Kay and Yoder 1987). For sources of information on each culture group's use of medicinal plants in this data assemblage, see the bibliographic essay.

2. See Kay 1976, 1977a, 1980b, 1982; Kay and Yoder 1987.

3. See Quezada 1977; Lavrin 1989.

BIBLIOGRAPHIC ESSAY

In part 2, the information on modern uses of medicinal plants derives from a few definitive sources for each culture group. To avoid unnecessary and distracting repetition there, I have chosen to cite the researchers individually and in more detail in this bibliographic essay. Much of the information on Mexican Americans in Arizona and New Mexico comes from my own research, or that of my students. Descriptions of uses of medicinal plants by Mexican Americans in Colorado was collected by botanists there, particularly Robert Bye, Jr.

The Pima Alto, who live in the Sonoran Desert of Arizona and Sonora, include the Tohono O'odham (formerly known as Papago) and Pima. Aleš Hrdlička listed many medicinal plant uses of the Tohono O'odham in *Physiological and Medical Observations among the Indians of the Southwestern United States and Northern Mexico* (1908). Another study of their traditional medicines can be found in Edward Castetter and Ruth Underhill's *Ethnobiology of the Papago Indians* (1935). Pima medicinal plant use was described by L.M.S. Curtin in *By the Prophet of the Earth* (1949). In addition, Gary Nabhan in *Gathering the Desert* (1985) has described the medicinal plants most valued by the O'odham: creosote bush and mesquite.

Farther south in Sonora, Campbell Pennington studied the medicinal plants of the Pima Bajo comprehensively, describing his findings in *The Pima Bajo of Central Sonora, Mexico* (1980). Earlier he had examined the traditional medicine of the Tepehuan (1969), the southernmost of the peoples of the American and Mexican West. In addition, he conducted research among the Mountain Pima (1973), who live in the vicinity of Yepáchic. More recently, Joseph Laferrière conducted an extensive study of Mountain Pima plant use (Laferrière et al. 1991; Laferrière n.d.).

Medicinal plant use by the Tarahumara—neighbors of the Tepehuan to the north, in Chihuahua—has been studied extensively by many ethnobotanists, perhaps in part because the Tarahumara are noted for collecting medicinal plants, which they sell throughout Mexico as well as in Ciudad Juarez, near the U.S.-Mexico border. *The Tarahumara* (1935), by Wendell Bennett and Robert Zingg, served as the foundation for the ethnobotanical literature of this people. More recent scholars include Pennington (1963) and Robert A. Bye, Jr. (1976, 1979, 1986).

West of the Tarahumara live the Warijio, who were until recently isolated from the surrounding Mexican culture. Howard Gentry gathered information about their ethnobotany, which he disseminated in *Rio Mayo Plants* (1942) and *The Warihio Indians of*

Sonora-Chihuahua (1962). Richard Felger and his associates are updating Gentry's classic work.

The Mayo, southwest of the Warijio, no longer live in the same valleys as they once did, but they have nonetheless retained much of their traditional plant pharmacopoeia, which shows more Mexican influence than do the pharmacopoeias of other indigenous groups. Information on medicinal plant use by the Mayo comes from data collected in the *municipio* of Masiaca by Mary O'Connor (Kay and O'Connor 1988) and in Las Animas, near Huatabampo, by Ismael Cozarit (1985). O'Connor has continued to interview the Mayo.

The Yaqui, who once lived to the north of the Mayo, have been studied intensively by Edward Spicer and his students for fifty years (Spicer 1980, 1983). The earliest of these ethnobotanies ("Medical Practices of the Yaquis"), written by Charles Wagner, was published in 1936. More recent ethnobotanists include Mary Elizabeth Shutler in Arizona (1967) and Anita Alvarez de Williams and Richard Felger in Sonora.

Still farther north, the Seri live in an extremely arid coastal environment from which they collect most of their medicinal plants. Richard Felger and Mary Moser comprehensively described the Seri ethnobotany in the now-classic *People of the Desert and Sea* (1985); Moser (1982) also provided an extensive obstetrical ethnobotany for these isolated people.

Across the Gulf of California at the northern end of Baja California Norte, the Paipai remained largely independent of the Mexican government but nevertheless borrowed many of their medicinal plants from the European-Mexican plant pharmacopoeia. Roger Owen (1963) collected information on herbal treatments used by the inhabitants of Santa Catarina in 1958–59.

The rest of Baja California also has not gone unnoticed. Recently, Michael Winkelman reported interviews with herbalists in Baja California Norte (1986) and treatments for diabetes (1989), while at the southernmost end of the peninsula, Rosalba Encarnación Dimayuga and colleagues (Encarnación and Agundez 1986; Encarnación, Fort, and Luis 1987; Encarnación and Keer 1991) interviewed elderly informants regarding medicinal plants in Baja California Sur.

Mexico has had a sustained interest in the study of traditional plant medicine, as evidenced by the establishment of IMEPLAN, the Mexican Institute for the Study of Medicinal Plants. Xavier Lozoya and his colleagues have extensively examined the traditional plant medicines of Mexico (Lozoya 1980, 1990; Lozoya and Lozoya 1982; Lozoya, Velázquez, and Flores 1988). New studies are reported regularly in the Mexican literature.

The first notable description of the folk medicine of Mexican Americans, *Cultural Differences and Medical Care: The Case of the Spanish-Speaking People of the Southwest* (1954), was made by Lyle Saunders. It was followed by a model of medical ethnography, Margaret Clark's (1959) *Health in the Mexican-American Culture*. Arthur Rubel contributed many works, in particular, "Concept of Disease in Mexican-American Culture" (1960). Karen Cowan Ford provided us with *Las Yerbas de la Gente: A Study of Hispano-American Medicinal Plants* (1975).

Other works on the plant pharmacopoeias of Mexican Americans are more narrowly focused by region or subject. William Madsen (1964) included analysis of folk medicine in his study of South Texas, Claudia Madsen (1968) studied its change, and Shirley Achor (1978) looked at its role in a Texas barrio. Robert Trotter and his *curandero* colleague Juan Chavira have extensively studied the folk medical practices of Mexican Americans in South Texas, giving special attention to plant remedies (Trotter 1981a, 1981b; Trotter and Chavira 1981). I researched a Mexican barrio known pseudonymously as El Jardín

(1972, 1977a), constructing a model of illness from the viewpoint of women. In *Healing Herbs of the Upper Rio Grande* (1965), L.M.S. Curtin reported the plants used as medicines of Spanish Americans in New Mexico. Robert Bye's comparison between Tarahumara and Mexican medicinals (1986) was particularly useful, as was "Ethnobotanical Notes from the Valley of San Luis, Colorado" (Bye and Linares 1986). Isabel Kelly, in *Folk Practices in North Mexico* (1965), described some plants used in folk medicine and spiritualism in the desert Mexican state of Coahuila in 1953.

Important work has been done on "women's conditions." I have looked at childbirth practices (Kay 1980b) and women's ethnotherapeutics (Kay and Yoder 1987). George Conway and John Slocumb produced a similarly focused work, "Plants Used as Abortifacients and Emmenagogues by Spanish New Mexicans" (Conway and Slocumb 1979). Sister M. Lucia van der Eerden examined maternity care in New Mexico (1948).

Many of the current writers interested in the disease theory held by Mexicans and Mexican Americans refer to George Foster's (1953) study of Spanish origins of New World folk medicine. Even today, Andrew Weil (see foreword) supports the idea of an ancient, unitary Old World source. However, arguing against Foster's thesis of imposed Spanish ideas, Alfredo López Austin (1970, 1980) and Bernard Ortiz de Montellano (1990) find that indigenous theories persist in folk medicine and that these indigenous concepts paved the way for absorbing European theory to form contemporary ethnic medicine. Carmen Anzures y Bolaños (1983) has written comprehensively on traditional medicine, the historical process, and syncretisms. My own work suggests that the similarities in ethnomedical theories are based on common human physiology.

The ethnobotanists listed were in most cases careful to deposit voucher specimens together with scientific names for plants along with their common names and descriptions. But the medicinal plant that had not been identified by a qualified botanist offered a challenge. When in doubt, I first referred to Ford's (1975) compendium if the plant listed therein had been properly identified. I compared information from Thomas Kearney and Robert Peebles's *Flowering Plants and Ferns of Arizona* (1942) and Norman Roberts's *Baja California Plant Field Guide* (1989). Next I tried the reference used in Mexican *yerberias*, Maximinio Martínez's *Las plantas medicinales de Mexico* (1969). I also used Stephen White's (1948) description of the vegetation and flora of the region of the Rio de Bavispe in northeastern Sonora; and Martínez's *Catalogo de nombres vulgares y científicos de plantas mexicanas* (1979) provided many possibilities. Also helpful was a mimeographed document, *Vegetacion del Estado de Sonora* (1972), made by the Sub-Agencia de Agricultura y Ganaderia in Hermosillo, Sonora, and given to me by Gary Nabhan, who was always helpful. José Luis Diaz's (1976–77) index of Mexican medicinal plants for IMEPLAN provided additional verifications. Pio Font Quer's *Plantas medicinales: El Dioscórides renovado* (1979) was authoritative for contemporary Spain.

I also used many dictionaries, first Francisco Santamaria's *Diccionario de mejicanismos* (1978), although it contains some errors. Horacio Sobarzo's *Vocabulario sonorense* (1966) provided some etymological information and was especially helpful with the most difficult identifications for medicinal plants of the eighteenth century. Howard Collard and Elizabeth Scott Collard's *Castellano-Mayo, Mayo-Castellano* dictionary of the Summer Institute of Linguistics (1962) also supplied names. The *Diccionario de la lengua española* (1970), *Pequeño Larousse Ilustrado* (1964), the unabridged *Larousse Spanish-English, English-Spanish Dictionary* (1993), and *The Compact Edition of the Oxford English Dictionary* (1971) were other sources.

I recommend that readers interested in seeing the medicinal plants described in part 2 visit botanical gardens such as the Arizona-Sonora Desert Museum west of Tucson;

the Desert Botanical Garden in Papago Park, Phoenix; and the Boyce Thompson Southwestern Arboretum, by Superior, Arizona. The plants in American botanical gardens are labeled by species and family names but do not tell of specific medicinal uses, unlike the detailed signs in European gardens.

Ethnomedicine is a dynamic field: there is fresh information about known plants and newly investigated plants. I have recorded only a few of the chemicals that have been identified by phytochemists in each plant. Interested readers should consult Duke 1985 and 1992, or access his database at World Wide Web, currently http://www.ars-grin.gov/~ngrlsbi

Information may be obtained from NAPRALERT (Natural Products Alert), a data base maintained by the Program for Collaborative Research in the Pharmaceutical Sciences, within the Department of Medicinal Chemistry and Pharmacognosy in the College of Pharmacy of the University of Illinois at Chicago, 833 South Wood Street, Chicago, Illinois 60612.

The *Journal of Ethnopharmacology* is a scientific publication that presents peer-reviewed international research. *HerbalGram,* also peer-reviewed but directed toward a less specialized reader, is an especially useful journal for recent information. The American Botanical Council, which publishes *HerbalGram,* maintains a bookstore with a large selection of scientific works including videos on botany, ethnobotany, pharmacognosy, the recognition and study of natural products, herbals from various cultures, and monographs. These may be ordered from American Botanical Council, P.O. Box 201660, Austin, Texas 78720.

REFERENCES

Abelson, Philip H.
    1990   Medicine from Plants. *Science* 247:513.
Achor, Shirley
    1978   *Mexican Americans in a Dallas Barrio.* Tucson: University of Arizona Press.
Acosta, Joseph de
    1962   *Historia natural y moral de las Indias.* 2nd ed. Edited by E. O'Gorman. Mexico City: Fondo de Cultura Economica. Originally published 1590.
Aguilar Contreras, Abigail, and Carlos Zolla
    1982   *Plantas tóxicas de México.* Mexico City: Instituto Mexicano del Seguro Social.
Aguirre Beltran, Gonzalo
    1947   La medicina indígena. *América Indígena* 7:107–27.
    1963   *Medicina y magia: El proceso de aculturación en la estructura colonial.* Mexico City: Instituto Nacional Indigenista.
Alegre, Francisco Javier
    1956   *Historia de la Compania de Jesus de Nueva España.* Rome: Institutum Historicum.
Almstedt, Ruth Farrell
    1977   *Digueño Curing Practices.* San Diego Museum Papers no. 10. San Diego: Museum of Man.
Amarillas, Joaquin de
    1783   Informe escrita desde estio de Xiaqui. WMS Amer. 50. Wellcome Institute for the History of Medicine, London.
Anderson, Eugene Newton
    1987   Why Is Humoral Medicine So Popular? *Social Science and Medicine* 25:331–37.
    1988   *The Food of China.* New Haven: Yale University Press.
Anderson, Robert
    1991   The Efficacy of Ethnomedicine: Research Methods in Trouble. *Medical Anthropology* 13:1–17.
Anonymous
    1749   Receta de las virtudes de las jojobas. WMS Amer. 72. Wellcome Institute for the History of Medicine, London.

Anzures y Bolaños, Carmen
    1983   *La medicina tradicional en México.* Mexico City: UNAM.
Appelt, Glenn D.
    1985   Pharmacological Aspects of Selected Herbs Employed in Hispanic Folk Medicine in the San Luis Valley of Colorado, USA: I. *Ligusticum porteri* (OSHA) and *Matricaria chamomilla* (Manzanilla). *Journal of Ethnopharmacology* 13:51–55.
Arano, Luisa Cogliati
    1976   *Tacuinum Sanitatis: The Medieval Health Handbook.* Translated and adapted by Oscar Ratti and Adele Westbrook. New York: George Braziller.
Ayensu, Edward S.
    1979   Plants for Medicinal Uses with Special Reference to Arid Zones. In *Arid Land Plant Resources: Proceedings of the International Arid Lands Conference on Plant Resources,* ed. J. R. Goodin and D. K. Northington. Lubbock, Texas: International Center for Arid and Semi-Arid Land Studies.
Bah, Moustapha, Robert Bye, and Rogelio Pereda-Miranda
    1994   Hepatotoxic Pyrrolizidine Alkaloids in the Mexican Medicinal Plant *Packera candidissima* (Asteraceae: Senecioneae). *Journal of Ethnopharmacology* 43: 19–30.
Balick, Michael J.
    1990   Ethnobotany and the Identification of Therapeutic Agents from the Rainforest. In *Bioactive Compounds from Plants: Ciba Foundation Symposium 154,* ed. D. J. Chadwick and J. Marsh. Chichester: John Wiley and Sons.
Barco, Miguel del
    1973   *Historia natural y cronica de la antigua California.* Edited by Miguel Leon Portilla. Mexico City: UNAM, Instituto de Investigaciones Historicas. Originally published 1768.
Barrios, Juan
    1607   *Verdaderamedicina, cirugia, y astrologia entre libros dividos por el doctor Inoan de Barrios: Natural de colmenar viejo.* Mexico City: Fernando Balli.
Bastien, Joseph
    1987   *Healers of the Andes.* Salt Lake City: University of Utah Press.
Beals, Ralph L.
    1932   The Comparative Ethnology of Northern Mexico before 1750. *Ibero-Americana* 2:93–225.
    1953   Acculturation. In *Anthropology Today,* ed. A. L. Kroeber, p. 621. Chicago: University of Chicago Press.
Begley, Sharon, with Elizabeth Ann Leonard
    1992   Take Two Roots; Call Me . . . *Newsweek* (February 3):53–54.
Bender, George
    1983   Searching the Southwest for Medicines. *Journal of Arizona History* 24:103–18.
Bennett, Wendell C., and Robert M. Zingg
    1935   *The Tarahumara: An Indian Tribe of Northern Mexico.* Chicago: University of Chicago Press.
Bicudo de Almeida, L., and M. Penteado
    1987   Carotenoids and Provitamin A Value of *Arracacia xanthorrhiza* Bancr. *Rev. Farm. Bioquim.* Sao Paulo, Brazil. Chemline Abstracts.
Biotherapeutic Index
    1993   Biologiche Heimittel. Baden-Baden, Germany: Heel GmisH.

Blumenthal, Mark
   1993a   Reports on Regulatory Dilemma. *HerbalGram* 28:11–12.
   1993b   Herb Industry and FDA Issue Chaparral Warning. *HerbalGram* 28:38–39, 53, 63, 69.
Boericke, William
   1994   *Pocket Manual of Homoeopathic Materia Medica and Repertory.* Reprint ed. New Delhi, India: B Jain Publishers.
Bourke, D. O'D., L. Fanjul, and A. J. Rendell-Dunn
   1987   *Spanish-English Horticultural Dictionary.* Farnham House, U.K.: C.A.B. International.
Brinker, Francis
   1991–   *Artemisia tridentata* Nutt. (Big Sagebrush). *British Journal of Phytotherapy*
   1992   2:97–114.
Brown, David E.
   1982   Biotic Communities of the American Southwest—United States and Mexico. *Desert Plants* 4. Special issue.
Brown, Michael F.
   1987   Toward a Human Ecology: Medical Ethnobotany and the Search for Dynamic Models of Plant Use. *Reviews in Anthropology* 14:5–11.
Browner, Carole H.
   1991   Gender Politics and Herbal Knowledge. *Medical Anthropology Quarterly* 5:99–124.
Browner, Carole H., Bernard R. Ortiz de Montellano, and Arthur J. Rubel
   1988   A Methodology for Cross-Cultural Ethnomedical Research. *Current Anthropology* 29:681–702.
Burton, Robert
   1927   *The Anatomy of Melancholy.* Translated by Floyd Dell and Paul Jordan-Smith. New York: Tudor Publishing. Originally published 1620.
Bye, Robert A., Jr.
   1976   *The Ethnoecology of the Tarahumara of Chihuahua, Mexico.* Ph.D. diss., Harvard University.
   1979   Hallucinogenic Plants of the Tarahumara. *Journal of Ethnopharmacology* 1:23–48.
   1986   Comparative Study of Tarahumara and Mexican Market Plants. *Economic Botany* 40:103–24.
Bye, Robert A., Jr., and Edelmira Linares
   1986   Ethnobotanical Notes from the Valley of San Luis, Colorado. *Journal of Ethnobiology* 6, no. 2:289–306.
Bye, Robert A., Jr., Edelmira Linares, Rachel Mata, Carlos Albor, Perla Castañeda, and Guillermo Delgado
   1991   Ethnobotanical and Phytochemical Investigation of *Randia Echinocarpa* (Rubiaceae). *Anales Instituto Biologico Universidad Nacional Autónoma México, Serie Botánico* 62:87–106.
Bye, Robert A., Jr., Rachel Mata, and José Pimentel
   1991   Botany, Ethnobotany and Chemistry of *Datura Lanulosa* (Solanaceae) in Mexico. *Anales Instituto Biologico Universidad Nacional Autónoma México, Serie Botánico* 61:21–42.
Cáceres, Armando, Orlando Cano, Blanca Samayoa, and Leila Aguilar
   1990   Plants Used in Guatemala for the Treatment of GI Disorders. 1: Screening of 84 Plants against Enterobacteria. *Journal of Ethnopharmacology* 30:55–73.

Cáceres, Armando, Alma V. Alvarez, Ana E. Ovando, and Blanca E. Samayoa
    1991   Plants Used in Guatemala for the Treatment of Respiratory Diseases. 1: Screening of 68 Plants against Gram-Positive Bacteria. *Journal of Ethnopharmacology* 31:193–208.

Cáceres, Armando, Ligia Fletes, Leila Aguilar, Olvi Ramirez, Ligia Figueroa, Ana Maria Taracena, and Blanca Samayoa
    1993   Plants Used in Guatemala for the Treatment of Gastrointestinal Disorders. 3: Confirmation of Activity against Enterobacteria of 16 Plants. *Journal of Ethnopharmacology* 38:31–38.

Castetter, Edward F., and Ruth M. Underhill
    1935   *The Ethnobiology of the Papago Indians.* University of New Mexico Bulletin no. 275. Ethnobiological Series vol. 4, no. 3. Albuquerque.

Castillo de Lucas, Antonio
    1958   *Folkmedicina.* Madrid: Editorial Dossat.

Clark, Margaret
    1959   *Health in the Mexican-American Culture.* Berkeley: University of California Press.

Clarkson, Rosetta E.
    1972   *The Golden Age of Herbs and Herbalists.* New York: Dover Publications.

Clavijero, Francisco Xavier, S.J. [Clavigero, Francisco Saverino]
    1787   *The History of Mexico.* Translated from Italian by Charles Cullen. London: G. J. and J. Robinson.

    1937   *The History of Lower California.* Translated from Italian and edited by Sara Lake and A. A. Gray. Stanford: Stanford University Press. Originally published 1786.

Cobo, Padre Bernabe
    1890   *Historia natural.* Seville, Spain: D. Marcos Jiminez de la España. Originally published 1653.

Collard, Howard, and Elizabeth Scott Collard
    1962   *Castellano-Mayo, Mayo-Castellano.* Mexico City: Instituto Linguistico de Verano, with the Secretaría de Educación Pública.

Comas, Juan
    1954   Influencia indígena en la medicina hipocrática, en la Nueva España del siglo XVI. *América Indígena* 14:327–61.

    1964   Un caso de aculuracíon farmacológica en la Nueva España del siglo XVI: El tesoro de medicinas de Gregorio López. *Anales de Antropología* 1:145–73.

    1968   La medicina aborigen mexicana en la obra de Fray Augustín de Vetancourt (1698). *Anales de Antropología* 5:129–62.

    1971   Influencia de la farmacopea y terapéutica indígenas de Nueva España en la obra de Juan de Barrios (1607). *Anales de Antropología* 8:125–50.

Conway, George A., and Slocumb, John C.
    1979   Plants Used as Abortifacients and Emmenagogues by Spanish New Mexicans. *Journal of Ethnopharmacology* 1:241–61.

Coombes, Allen J.
    1985   *Dictionary of Plant Names.* Portland, Oreg.: Timber Press.

Cozarit M., Ismael
    1985   *Medicina tradicional Mayo.* Cuadernos de trabajo no. 2. Mimeo. Hermosillo: Unidad Regional Sonora Dirección General de Culturas Populares SEP.

Croke, Alexander
  1830   *Regimen Sanitatis Salernitatum.* London: Helme and Busby. Originally published 1607.
Croom, Edward M., Jr.
  1983   Documenting and Evaluating Herbal Remedies. *Economic Botany* 37:13–27.
Curtin, Leonora Scott Muse
  1949   *By the Prophet of the Earth.* Santa Fe, N.Mex.: San Vicente Foundation.
  1965   *Healing Herbs of the Upper Rio Grande.* Los Angeles: Southwest Museum.
Dean, Glenna
  1993   Use of Pollen Concentrations in Coprolite Analysis: An Archaeobotanical Viewpoint with a Comment to Reinhard et al. *Journal of Ethnobotany* 13:102–14.
Delgado, Guillermo, and José Garduno
  1987   Pyranocoumarins from *Arracacia nelsonii. Phytochemistry* 26:1139–41. Chemline Abstracts.
Der Marderosian, Ara, and Lawrence Liberti
  1988   *Natural Product Medicine.* Philadelphia: George F. Stickley.
Diaz, José Luis (editor)
  1976–   *Índice y sinonimia de las plantas medicinales de Mexico.* Mexico City:
  1977   IMEPLAN.
Dillehay, Tom D.
  1987   By the Banks of the Chinchihuapi. *Natural History* 90, no. 4:9–11.
Dioscorides, Pedanius of Anazarbos
  1959   *The Greek Herbal of Dioscorides: Englished by John Goodyer 1655.* Edited by R. T. Gunther. 1933. New York: Hafner Publishing.
Dorland's Medical Dictionary
  1974   Philadelphia: W. B. Saunders.
Duke, James A.
  1985   *CRC Handbook of Medicinal Herbs.* Boca Raton, Fla.: CRC Press.
  1986   *Handbook of Northeastern Indian Medicinal Plants.* Lincoln, Neb.: Quarterman Publishers.
  1992   *Handbook of Phytochemical Constituents of GRAS Herbs and Other Economic Plants.* Boca Raton, Fla.: CRC Press.
Dunn, Oliver, and James F. Kelly
  1989   *The Diario of Christopher Columbus's First Voyage to America 1492–1493.* Oklahoma City: University of Oklahoma Press.
Dunne, Peter Masten, S.J.
  1940   *Pioneer Black Robes on the West Coast.* Berkeley and Los Angeles: University of California Press.
Eisenberg, D. M., R. C. Kessler, C. Foster, F. E. Norlock, D. R. Calkins, and T. L. Delbanco
  1993   Unconventional Medicine in the United States. *New England Journal of Medicine* 328:246–52.
Ellis, John
  1770   Directions for Bringing over seeds and plants from the East Indies and Other Distant Countries in a State of Vegetation together with a Catalogue of such Foreign Plants as are worthy of being encouraged in our American Colonies for the purposes of Medicine, Agriculture and Commerce. Wellcome document 21631. Wellcome Institute for the History of Medicine, London.

Emmart, Emily Walcott

1940   *The Badianus Manuscript.* Baltimore: Johns Hopkins University Press.

Encarnación, Rosalba Dimayuga, and Jorge Agundez

1986   Traditional Medicine of Baja California Sur (Mexico), I. *Journal of Ethnopharmacology* 17:183–93.

Encarnación, Rosalba Dimayuga, R. M. Fort, and Maritza Luis Pantoja

1987   Traditional Medicine of Baja California Sur (Mexico), II. *Journal of Ethnopharmacology* 20:209–22.

Encarnación, Rosalba Dimayuga, and Sergio Keer Garcia

1991   Antimicrobial Screening of Medicinal Plants from Baja California Sur, Mexico. *Journal of Ethnopharmacology* 31:181–92.

Engstrand, Iris H. W.

1981   *Spanish Scientists in the New World: The Eighteenth Century Expeditions.* Seattle: University of Washington Press.

Esteyneffer, Juan de

1887   *Florilegio medicinal de todas las enfermedades o breve epidome de las medicina y cirujia la primera obra sobre esta ciencia impresa en Mexico en 1713.* 100 vols. Mexico City: Biblioteca Mexicana.

1978   *Florilegio medicinal de todas las enfermedades.* Edited by Carmen Anzures de Bolaños. Mexico City: Academia Nacional de Medicina. Originally published 1719.

Etkin, Nina L.

1986   Multidisciplinary Perspectives in the Interpretation of Plants. In *Plants in Indigenous Medicine and Diet.* Bedford Hills, N.Y.: Redgrave.

1988   Ethnopharmacology: Biobehavioral Approaches in the Anthropological Study of Indigenous Medicines. *Annual Review of Anthropology* 17:23–42.

Etkin, Nina L. (editor)

1994   *Eating on the Wild Side: The Pharmacologic, Ecologic, and Social Implications of Using Noncultigens.* Tucson: University of Arizona Press.

Farfán, Agustín

1944   *Tractado breve de medicina.* Colección de Incunales Americanos, vol. 10. Madrid: Ediciones Cultura Hispana. Originally published 1592.

Farnsworth, Norman R.

1983   The NAPRALERT Data Base as an Information Source for Application to Traditional Medicine. In *Traditional Medicine and Health Care Coverage,* ed. R. Bannerman, H. Obuasa, J. Burton, and W.-C. Chen, pp. 184–93. Geneva: World Health Organization.

1993a   Relative Safety of Herbal Medicines. *HerbalGram* 29:Special Supplement 36A–H.

1993b   Ethnopharmacology and Future Drug Development: The North American Experience. *Journal of Ethnopharmacology* 38:145–52.

Felger, Richard S., and Mary B. Moser

1985   *People of the Desert and Sea.* Tucson: University of Arizona Press.

Finkler, Kaja

1983   Studying Outcomes of Mexican Spiritualist Therapy. In *The Anthropology of Medicine: From Culture to Method,* ed. L. Romanucci-Ross, D. E. Moerman, and L. R. Tancredi. 1st ed. New York: Bergrin and Garvey.

Fontana, Bernard L.

1983   Pima and Papago: Introduction. In *Handbook of North American Indians,* vol. 10, vol. ed. A. Ortiz. Washington, D.C.: Smithsonian Institution Press.

Font Quer, Pio
  1979   *Plantas medicinales: El Dioscorides renovado.* 5th ed. Barcelona: Editorial Labor.

Ford, Karen Cowan
  1975   *Las Yerbas de la Gente: A Study of Hispano-American Medicinal Plants.* University of Michigan Museum of Anthropology, Anthropological Papers no. 60. Ann Arbor.

Fortuine, Robert
  1988   The Use of Medicinal Plants by the Alaska Natives. *Alaska Medicine* 30:185–226.
  1992   *Chills and Fever: Health and Disease in the Early History of Alaska.* Fairbanks: University of Alaska Press.

Foster, George M.
  1953   Relationships between Spanish and Spanish-American Folk Medicine. *Journal of American Folklore* 66:201–17.
  1960   *Culture and Conquest.* Viking Fund Publications in Anthropology no. 27. Chicago: Quadrangle Books.
  1988   The Validating Role of Humoral Theory in Traditional Spanish-American Therapeutics. *American Ethnologist* 15:120–35.
  1994   *Hippocrates' Latin American Legacy: Humoral Medicine in the New World.* Amsterdam: Gordon and Breach Science Publishers.

Foster, Steven, and James A. Duke
  1990   *A Field Guide to Medicinal Plants.* Boston: Houghton Mifflin.

Fuller, Thomas C., and Elizabeth McClintock
  1986   *Poisonous Plants of California.* Berkeley: University of California Press.

Gates, William
  1939   *The de la Cruz Badiano: Aztec Herbal of 1552.* Baltimore: Maya Society.

Gentry, Howard
  1942   *Rio Mayo Plants.* Carnegie Institution of Washington, Publication no. 527. Washington, D.C.
  1962   *The Warihio Indians of Sonora-Chihuahua: An Ethnographic Survey.* Bureau of American Ethnology, Anthropological Paper no. 65, Bulletin 186. Washington, D.C.

Gerard, John
  1597   *The Herball or Generall Historie of Plantes.* London: John Norton.
  1975   *The Herbal or General History of Plants: The Complete 1633 Edition as Revised and Enlarged by Thomas Johnson.* New York: Dover Publications.

Gicklhorn, Renée
  1973   *Missionsapotheker: Deutsche Pharmazeuten im Lateinamerika des 17. und 18. Jahrhunderts.* Stuttgart: Wissenschaftliche Verlagsgesellschaft, MBH.

Gilij, Filippo Salvadore
  1785   Remedios singulares usado por los misioneros de tierra firme. Vol. 3 of the manuscript collection Roma año 1782. Wellcome Institute for the History of Medicine, London.

Good, Ronald D'Oyley
  1974   *The Geography of the Flowering Plants.* 4th ed. London: Longman Group.

Graham, Frances Kelso
  1985   *Plant Lore of an Alaskan Island.* Anchorage: Alaska Northwest Publishing.

Granjel, Luis S.
  1965   *Historia de la pediatria española: Cuadernos de historia de la medicina*

española monografías III. Salamanca, Spain: Universidad de Salamanca, Graficas Europa, Sanchez Llevot.

1978   *La medicina española del siglo XVII: Historia general de la medicina española III*. Salamanca, Spain: Universidad de Salamanca, Graficas Europa, Sanchez Llevot.

1979   *La medicina española del siglo XVIII: Historia general de la medicina española IV*. Salamanca, Spain: Universidad de Salamanca, Graficas Europa, Sanchez Llevot.

Greenberg, J. H., and M. Ruhlen

1992   Linguistic Origins of Native Americans. *Scientific American* (November): 94–99.

Greenberg, J. H., C. G. Turner, and S. L. Zegura

1986   The Settlement of the Americas: A Comparison of the Linguistic, Dental, and Genetic Evidence. *Current Anthropology* 27:477–97.

Hammond, George P., and Agapito Rey

1953   *Don Juan de Oñate, Colonizer of New Mexico*. Albuquerque: University of New Mexico Press.

Harbottle, Garman, and Phil C. Weigand

1992   Turquoise in Pre-Columbian America. *Scientific American* (February): 78–85.

Hedrick, V. P. (editor)

1972   *Sturtevant's Edible Plants of the World*. New York: Dover Press.

Heiser, Robert F., and Albert B. Elsasser

1980   *The Natural World of the California Indians*. Berkeley: University of California Press.

Hernández, Francisco

1942   *Historia de las plantas de Nueva España*. Mexico City: Universidad Nacional Autonoma, Instituto de Biologia Imprenta Universitaria.

1959   *Historia natural de Nueva España*. Vols. 2 and 3 of *Obras completas*. Mexico City: UNAM.

1960   *Vida y obra de Francisco Hernández*. Vol. 1 of *Obras completas*. Mexico City: UNAM.

1966   *Historia natural de Cayo Plinio Segundo*. Vol. 4 of *Obras completas*. Mexico City: UNAM.

1976   *Historia natural de Cayo Plinio Segundo*. Vols. 5 and 5ii of *Obras completas*. Mexico City: UNAM.

1976   *Antigüedades de la Nueva España*. Vol. 6 of *Obras completas*. Mexico City: UNAM.

Herrera, Maria Teresa

1973   *Menor daño de la medicina de Alonso de Chirino (d. 1429)*. Salamanca, Spain: Ed. Critica.

Hinton, Thomas B.

1983   Southern Periphery: West. In *Handbook of North American Indians*, vol. 10, vol. ed. A. Ortiz. Washington, D.C.: Smithsonian Institution Press.

Houghton, P. J.

1984   Ethnopharmacology of Some Buddleja Species. *Journal of Ethnopharmacology* 11:293–308.

Hrdlička, Aleš

1904   The Indians of Sonora, Mexico. *American Anthropologist* 6:51–89.

1908   *Physiological and Medical Observations among the Indians of Southwestern United States and Northern Mexico.* Bureau of American Ethnology, Bulletin 34. Washington, D.C.

Hutchens, Alma R.
1973   *Indian Herbalogy of North America.* Windsor, Ontario: Merco.

Huxtable, Ryan J.
1980   Herbal Teas and Toxins: Novel Aspects of Pyrrolizidine Poisoning in the U.S. *Perspectives in Biology and Medicine* 24:1–13.

1983   Herbs along the Western Mexican-American Border. *Proceedings of the Western Pharmacological Society* 26:185–91.

1990   The Harmful Potential of Herbal and Other Plant Products. *Drug Safety* 5, Supp. 1:126–36.

1992   The Pharmacology of Extinction. *Journal of Ethnopharmacology* 37:1–11.

Jackson, Benjamin D.
1965   *A Life of William Turner (1877): In Facsimile Edition of Libellus de Re Herbaria [1538] and The Names of Herbes [1548].* London: Ray Society.

Janzen, Daniel H.
1978   Complications in Interpreting the Chemical Defenses of Trees against Tropical Arboreal Plant-Eating Vertebrates. In *The Ecology of Arboreal Folivores,* ed. G. G. Montgomery. Washington, D.C.: Smithsonian Institution Press.

Johns, Timothy
1990   *With Bitter Herbs They Shall Eat It: Chemical Ecology and the Origins of Human Diet and Medicine.* Tucson: University of Arizona Press.

Johnson, Matthew B.
1992   The Genus *Bursera* (Burseraceae) in Sonora, Mexico, and Arizona, U.S.A. *Desert Plants* 10:126–43.

Kaempfer, Engelbert
1906   *History of Japan.* 3 vols. Translated by J. G. Schenchzer. Glasgow: J. MacLehose and Sons. Originally published 1727.

Kari, Priscilla Russell
1991   *Tanaina Plantlore, Dena'ina K'et'una.* 3rd ed. Fairbanks: Alaska Native Language Center.

Kay, Margarita Artschwager
1972   *Health and Illness in the Barrio: Women's Point of View.* Ph.D. diss., University of Arizona.

1974   The Ethnosemantics of Mexican American Fertility. American Anthropological Association annual meetings, Mexico City.

1976   The Fusion of Utoaztecan and European Ethnogynecology in the *Florilegio Medicinal.* In *Actas del XLI Congreso Internacional de Americanistas,* vol. 3. Mexico City: Instituto Nacional de Antropología e Historia.

1977a   Health and Illness in a Mexican Barrio. In *Ethnic Medicine in the Southwest,* ed. E. H. Spicer. Tucson: University of Arizona Press.

1977b   The Florilegio Medicinal: Source of Southwestern Ethnomedicine. *Ethnohistory* 28:251–59.

1978   Parallel, Alternative, or Collaborative: Curanderismo in Tucson, Arizona. In *Modern Medicine and Medical Anthropology in the United States–Mexico Border Population,* ed. B. Velimirovic. Scientific Publication no. 359. Washington, D.C.: Pan American Health Organization.

1979   Lexemic Change and Semantic Shift in Disease Names. *Culture, Medicine and Psychiatry* 3:73–94.

1980a   Poisoning by Gordolobo: Why Was the Wrong Herb Collected? American Anthropological Association annual meetings, Los Angeles.

1980b   Mexican, Mexican American, and Chicana Childbirth. In *Twice a Minority: Mexican American Women,* ed. M. B. Melville. St. Louis: C. V. Mosby.

1987   Lay Theory Healing in Northwestern New Spain. *Social Science and Medicine* 24:1051–60.

1993   Fallen Fontanelle: Culture-Bound or Cross-Cultural? *Medical Anthropology* 15:137–56.

1994   Poisoning by Gordolobo. *HerbalGram* 32:42–46, 57.

Kay, Margarita, and Mary K. O'Connor

1988   The Medical Ethnobotany of the Mayo. American Anthropological Association annual meetings, Phoenix.

Kay, Margarita, and Guadalupe Olivas

1982   Mexican American Grandmothers as Health Care Advisors. Unpublished data.

Kay, Margarita, Ann Voda, Guadalupe Olivas, Frances Rios, and Margaret Imle

1982   Ethnography of the Menopause-Related Hot Flash. *Maturitas* 4:217–27.

Kay, Margarita, and Marianne Yoder

1987   Hot and Cold in Women's Ethnotherapeutics: The American Mexican West. *Social Science and Medicine* 25:347–55.

Kearney, Thomas H., and Robert H. Peebles

1942   *Flowering Plants and Ferns of Arizona.* United States Department of Agriculture, Miscellaneous Publication no. 423. Washington, D.C.

Kelley, Bruce D., Glenn D. Appelt, and Jennifer M. Appelt

1988   Pharmacological Aspects of Selected Herbs Employed in Hispanic Folk Medicine in the San Luis Valley of Colorado, USA: II. *Asclepias asperula* (Inmortal) and *Achillea lanulosa* (Plumajillo). *Journal of Ethnopharmacology* 22:1–9.

Kelly, Isabel

1965   *Folk Practices in North Mexico.* Austin: University of Texas Press.

Kincaid, Jamaica

1992   Flowers of Evil. *New Yorker* (October 5):154–59.

Kingsbury, John M.

1964   *Poisonous Plants of the United States and Canada.* Englewood Cliffs, N.J.: Prentice-Hall.

Kirchhoff, Paul

1954   Gatherers and Farmers in the Greater Southwest: A Problem in Classification. *American Anthropologist* 56:529–50.

Kourennoff, Paul M.

1970   *Russian Folk Medicine.* Edited and translated by George St. George. London: W. H. Allen.

Krochmal, Arnold, and Connie Krochmal

1973   *A Guide to the Medicinal Plants of the United States.* New York: Quadrangle/ The New York Times Book Co.

Laferrière, Joseph E.

n.d.   Mountain Pima Medicinal Plants. Unpublished data.

Laferrière, Joseph E., C. W. Weber, and E. A. Kohlhepp

1991   Use and Nutritional Composition of Some Traditional Mountain Pima Plant Foods. *Journal of Ethnobiology* 11:93–114.

Lavrin, Asunción
1989   Sexuality in Colonial Mexico: A Church Dilemma. In *Sexuality and Marriage in Colonial Latin America,* ed. A. Lavrin. Lincoln: University of Nebraska Press.

León-Portilla, Miguel
1972   The Norteño Variety of Mexican Culture. In *Plural Society in the Southwest,* ed. E. H. Spicer and R. H. Thompson. New York: Interbook.

Leroi-Gourhan, Arlette
1974   The Flowers Found with Shanidar IV, a Neanderthal Burial in Iraq. *Science* 190:562–64.

Lévi-Strauss, Claude
1966   *The Savage Mind.* Chicago: University of Chicago Press.

Lewis, Walter H., and Memory P. F. Elvin-Lewis
1977   *Medical Botany.* New York: John Wiley and Sons.

Li Shih-Chen
1973   *Chinese Medicinal Herbs.* Translated by F. Porter Smith and G. A. Stuart. San Francisco: Georgetown Press.

Lietava, Jan
1992   Medicinal Plants in a Middle Paleolithic Grave Shanidar IV? *Journal of Ethnopharmacology* 35:263–66.

Linares Mazari, Maria Edelmira, Robert Bye, and Beatriz Flores Peñafiel
1984   *Tes curativos de Mexico.* Mexico City: Universidad Nacional Autonoma de México.

López, Gregorio
1672   *Tesoro de medicina para diversas enfermedades.* Mexico City: Francisco Rodriguez Lupercio. Originally published 1580–1589.

López Austin, Alfredo
1970   Ideas etiológicas en la medicina nahuatl. *Anuario Indigenista* 30:255–75.
1980   *Cuerpo humano e ideología.* Mexico City: Universidad Nacional Autonoma de México.

López de Villalobos, Francisco
1973   *El sumario de la medicina con un tratado de las pestiferas bubas.* Edited by María Teresa Herrera. Cuadernos de Historia de la Medicina Española, Monografias XXV. Salamanca: Instituto de Historia de la Medicina Española. Originally published 1499.

López Estudillo, Rigoberto, and Hinojosa Garcia, Alicia
1988   *Catalogo de plantas medicinales sonorenses.* Hermosillo: Universidad de Sonora.

Lozoya, Xavier
1980   Mexican Medicinal Plants Used for Treatment of Cardiovascular Diseases. *American Journal of Chinese Medicine* 8:86–95.
1990   An Overview of the System of Traditional Medicine Currently Practiced in Mexico. In *Economic and Medicinal Plant Research,* vol. 4, *Plants and Traditional Medicine,* ed. H. Wagner and N. R. Farnsworth. New York: Academic Press.

Lozoya, Xavier, and Mariana Lozoya
1982   *Flora medicinal de Mexico,* part 1, *Plantas indigenas.* Mexico City: Instituto Mexicano del Seguro Social.

Lozoya, Xavier, Georgina Velázquez, and Angel Flores

   1988   *La medicina tradicional en Mexico.* Experiencia del Programa IMSS-COPLAMAR 1982–1988. Mexico City: Instituto Mexicano del Seguro Social.

Lozoya, Xavier, and Carlos Zolla (editors)

   1986   *La medicina invisible.* Mexico City: Folios Ediciones.

MacFarlane, R. G.

   1962   The Reactions of Blood to Injury. In *General Pathology,* ed. Sir Howard Florey. Philadelphia: W. B. Saunders.

MacLeish, Archibald

   1963   *The Collected Poems of Archibald MacLeish.* Cambridge, Mass.: Riverside Press.

MacNutt, Francis Augustus (editor and translator)

   1908   *Letters of Cortes.* New York: Putnam.

Madsen, Claudia

   1968   *A Study of Change in Mexican Folk Medicine.* New Orleans: Middle American Research Institute, Tulane University.

Madsen, William

   1964   *The Mexican-Americans of South Texas.* New York: Holt, Rinehart, and Winston.

Martin, Debra L.

   1990   Patterns of Health and Disease: Stress Profiles for the Prehistoric Southwest. Ms. prepared for lecture series, The Organization and Evolution of Prehistoric Southwestern Society, Department of Anthropology, University of Arizona. Tucson.

Martínez, Maximinio

   1969   *Las plantas medicinales de Mexico.* 5th ed. Mexico City: Ediciones Botas.

   1979   *Catalogo de nombres vulgares y científicos de plantas mexicanas.* Mexico City: Fondo de Cultura Economica.

Martínez Bravo, Eugenio

   1984   Sangre de grado—Estudios de campo y en clinica humana. Proceedings Segundo Congreso Internacional de Medicina Tradicional y Folklorica, Cuernavaca.

Mason, Charles T., Jr., and Patricia B. Mason

   1987   *A Handbook of Mexican Roadside Flora.* Tucson: University of Arizona Press.

Mathien, Frances Joan, and Randall H. McGuire (editors)

   1986   *Ripples in the Chichimec Sea: New Considerations of Southwestern-Mesoamerican Interactions.* Carbondale and Edwardsville: Southern Illinois University Press.

Mayes, Vernon O., and Barbara Bayless Lacey

   1989   *Nanise': A Navajo Herbal.* Tsaile, Ariz.: Navajo Community College Press.

McGuire, Randall

   1980   The Mesoamerican Connection in the Southwest. *Kiva* 46:3–38.

McGuire, Randall, and Michael Schiffer

   1982   *Hohokam and Patayan: Prehistory of Southwestern Arizona.* New York: Academic Press.

Merbs, Charles F., Robert J. Miller, and Elizabeth S. Dyer Alcauskas (editors)

   1985   *Health and Disease in the Prehistoric Southwest.* Arizona State University Anthropological Research Papers. Tempe.

Meredith, John Dee
1982   *Changes in the Ethnic Medical System of the Hispanic Population of Casper, Wyoming.* Ph.D. diss., University of Arizona.
Miller, Wick R.
1983   Uto-Aztecan. In *Handbook of North American Indians,* vol. 10, vol. ed. A. Ortiz. Washington, D.C.: Smithsonian Institution Press.
Millspaugh, Charles F.
1974   *American Medicinal Plants.* New York: Dover Publications. Originally published 1892.
Minnis, Paul E.
1991   Famine Foods of the North American Desert Borderlands in Historical Context. *Journal of Ethnobiology* 11:231–56.
Mitich, Larry
1983   The Succulent Euphorbias: Poisonous and Medicinal. *Euphorbia Journal* 1:61–67.
Moerman, Daniel E.
1986   *Medicinal Plants of Native America.* 2 vols. University of Michigan Museum of Anthropology, Technical Reports no. 19. Ann Arbor.
1991   The Medicinal Flora of Native North America: An Analysis. *Journal of Ethnopharmacology* 31:1–42.
Monardes, Nicholas
1574   *La historia de las cosas que se traen de nuestras Indias Occidentales.*
1596   *Joyfull Newes Out of the Newe Founde Worlde: Englished by John Frampton.* London: E. Allde, by the assigne of Bonham Norton.
Moore, Michael
1979   *Medicinal Plants of the Mountain West.* Santa Fe: Museum of New Mexico Press.
1989   *Medicinal Plants of the Desert and Canyon West.* Santa Fe: Museum of New Mexico Press.
Morfi, Juan Agustin
1980   *Viaje de Indios y diario del Nuevo Mexico.* Edited by Vito Alessio Robles. Mexico City: Manuel Porrúa. Originally published 1792.
Morison, Samuel Eliot (editor and translator)
1963   *The Journals and Other Documents on the Life and Voyages of Christopher Columbus.* New York: Heritage Press.
Morton, Julia
1981   *Atlas of Medicinal Plants of Middle America.* Springfield, Ill.: Charles C. Thomas.
Moser, Mary Beck
1982   Seri: From Conception through Infancy. In *Anthropology of Human Birth,* ed. M. A. Kay. Philadelphia: F. A. Davis.
Muller, Mary Clay
n.d.   *A Preliminary Checklist of the Vascular Plants in Southeastern Alaska.* U.S. Department of Agriculture, Forest Service, Alaska Region.
Murray, Michael T., and Joseph E. Pizzorno
1991   *An Encyclopedia of Natural Medicine.* Rocklin, Calif.: Prima Publications.
Nabhan, Gary Paul
1985   *Gathering the Desert.* Tucson: University of Arizona Press.

Nabhan, Gary, J. W Berry, and C. W. Weber
  1980   Wild Beans of the Greater Southwest: *Phaseolus metcalfei* and *P. ritensis. Economic Botany* 34:68–85.
Nentuig, Juan [Nentvig, Johann; Nentvig, Juan]
  1977   *El Rudo Ensayo.* Edited by Margarita Nolasco Armas, Teresa Martinez Peñalosa, and America Flores. Mexico City: SEP/INAH. Originally published 1764.
  1980   *Rudo Ensayo: A Description of Sonora and Arizona in 1764.* Translated by Alberto Francisco Pradeau and Robert R. Rasmussen. Tucson: University of Arizona Press.
Niethammer, Carolyn
  1974   *American Indian Food and Lore.* New York: Collier Macmillan.
Norberga, Sh. B.
  1975   Methionine, Cystine, Lysine, and Tryptophan in Some Venezuelan Foods. *Arch. Latinoam. Nutr.* 17:111–16.
Nuñez, Angel
  1777   Carta escrita desde la Mission de Santa Maria de Baserac al Muy Reverendo Padre Fray Manuel Riezu . . . dignisimo Ministro Provincial de la Santa Provincia de Santiago de Xalixco. Unpublished manuscript, Mission San Xavier del Bac, Tucson.
Nuttall, Zelia
  1923   The Gardens of Ancient Mexico. *Annual Report of the Smithsonian Institution,* pp. 453–64. Washington, D.C.: U.S. Government Printing Office.
O'Connor, Mary
  1989   *Descendants of Totoliguoqui: Ethnicity and Economics in the Mayo Valley.* University of California Publications in Anthropology, vol. 19. Berkeley: University of California Press.
Orellano, Sandra L.
  1987   *Indian Medicine in Highland Guatemala: The Pre-Hispanic and Colonial Periods.* Albuquerque: University of New Mexico Press.
Ortiz de Montellano, Bernard R. [Bernardo]
  1976   ¿Una clasificación botánica entre los nahoas? In *Estado actual del conocimiento en plantas medicinales mexicanas,* ed. X. Lozoya. Mexico City: IMEPLAN.
  1990   *Aztec Medicine, Health, and Nutrition.* New Brunswick, N.J.: Rutgers University Press.
Ortiz de Montellano, Bernard R., and Carole H. Browner
  1985   Chemical Bases for Medicinal Plant Use in Oaxaca, Mexico. *Journal of Ethnopharmacology* 13:57–88.
Oviedo y Valdés, Gonzalo Fernández de
  1851–   *Historia general y natural de las Indias.* Edited by Jose Amador de los Rios.
  1855   Madrid: Imprenta de la Real Academia de la Historia. Originally published 1535–1557.
  1986   *Sumario de la natural historia de las Indias.* Edited by Manuel Ballesteros Gaibrois. Madrid: Hermanos García Noblejas. Originally published 1526.
Owen, Roger C.
  1963   The Use of Plants and Non-Magical Technique in Curing Illness among the Paipai, Santa Catarina, Baja California, México. *América Indígena* 23:319–45.
Pablos, Blas Antonio
  1784   Informe escrita desde Rio Chico, Sonora. WMS Amer. 74. Wellcome Institute for the History of Medicine, London.

Pachter, Lee (guest editor)
1993   Latino Folk Illness. *Medical Anthropology* 15:103–213. Special issue.
Padron, Francisco
1956   *El medico y el folklore.* San Luis Potosí, Mexico: Talleres Graficos de la Editorial Universitaria.
Pammel, L. H.
1992   *A Manual of Poisonous Plants.* Forestburgh, N.Y.: Lubrecht and Cramer.
Pennington, Campbell W.
1963   *The Tarahumar of Mexico: Their Environment and Material Culture.* Salt Lake City: University of Utah Press.
1969   *The Tepehuan of Chihuahua.* Salt Lake City: University of Utah Press.
1973   Plantas medicinales utilizadas por el pima montañés de Chihuahua. *América Indígena* 33:213–32.
1980   *The Pima Bajo of Central Sonora, Mexico,* vol. 1. Salt Lake City: University of Utah Press.
1983a   Tarahumara. In *Handbook of North American Indians,* vol. 10, vol. ed. A. Ortiz. Washington, D.C.: Smithsonian Institution Press.
1983b   Northern Tepehuan. In *Handbook of North American Indians,* vol. 10, vol. ed. A. Ortiz. Washington, D.C.: Smithsonian Institution Press.
Pérez, Cristina, and Claudia Anesini
1994   In Vitro Antibacterial Activity of Argentine Folk Medicinal Plants against *Salmonella typhi. Journal of Ethnopharmacology* 44:41–46.
Pérez de Ribas, Andrés
1944   *Triunfos de Nuestra Santa Fé entre gentes las mas bárbaras y fieras del Nuevo Orbe,* vols. 1–3. Mexico City: Editorial Layac. Originally published 1645.
Pérez G., R. M., A. Ocegueda Z., J. L. Muñoz L., J. G. Avila A., and W. W. Morrow
1984   A Study of the Hypoglucemic Effect of Some Mexican Plants. *Journal of Ethnopharmacology* 12:253–62.
Pérez G., R. M., Yescas Laguna, and Aleksander Walkowski
1985   Diuretic Activity of Mexican Equisetum. *Journal of Ethnopharmacology* 14:269–72.
Pfefferkorn, Ignatz [Ignaz, Ignacio]
1983   *Descripción de la Provincia de Sonora.* 2nd vol. Translated to Spanish from Treutlein's English translation of the original German by Armando Hopkins Durazo. Hermosillo: Gobierno del Estado de Sonora. Originally published 1795.
1989   *Sonora: A Description of the Province.* Translated by Theodore E. Treutlein. Tucson: University of Arizona Press. Originally published in 2 vols., 1794–1795.
Pliny, the Elder
1855   *The Natural History of Pliny.* 5 vols. Translated by J. Bostock and H. T. Riley. London: Henry G. Bohn. Originally published ca. A.D. 70.
1938   *Natural History.* 10 vols. Translated by W.H.S. Jones. Cambridge, Mass.: Loeb Classical Library, Harvard University Press. Originally published ca. A.D. 70.
Polo, Marco
1958   *The Travels of Marco Polo.* New York: Orion Press.
Quezada, Noemí
1975   Métodos anticonceptivos y abortivos tradicionales. *Anales de Antropología* 12:223–42.

1977   Creencias tradicionales sobre embarazo y parto. *Anales de Antropología* 14:307–26.

Radbill, Samuel X.

1974   Pediatrics. In *Medicine in Seventeenth Century England,* ed. A. G. Debus. Berkeley: University of California Press.

Ramírez, Carlos

1989   The Macrobotanical Remains. In *Monte Verde: A Late Pleistocene Settlement in Chile,* vol. 1, ed. T. D. Dillehay. Washington, D.C.: Smithsonian Institution Press.

Ramos-Elorduy de Conconi, Julieta, and Jose Manuel Pino Moreno

1988   The Utilization of Insects in the Empirical Medicine of Ancient Mexicans. *Journal of Ethnobiology* 8:195–202.

Reinhard, Karl J., J. Richard Ambler, and Magdalene McGuffie

1985   Diet and Disease at Dust Devil Cave. *American Antiquity* 50:819–824.

Reinhard, Karl J., Donny L. Hamilton, and Richard H. Helvy

1991   Use of Pollen Concentration in Paleopharmacology: Coprolite Evidence of Medicinal Plants. *Journal of Ethnobiology* 11:117–32.

Reveal, James L.

1992   *Gentle Conquest: The Botanical Discovery of North America.* Washington, D.C.: Starwood Publishing.

Risse, Guenter B.

1987   Medicine in New Spain. In *Medicine in the New World,* ed. R. L. Numbers. Knoxville: University of Tennessee Press.

Roberts, Norman C.

1989   *Baja California Plant Field Guide.* La Jolla, Calif.: Natural History Publishing Company.

Rodriguez, Eloy

1992   Fire Medicine. Unpublished Sacnas Community Science Lecture. Laboratorios de Investigación en Química de Plantas Medicinales y Tóxicas, University of California, Irvine.

Rohde, Eleanour Sinclair

1971   *The Old English Herbals.* New York: Dover. Originally published 1921.

Rubel, Arthur J.

1960   Concept of Disease in Mexican-American Culture. *American Anthropologist* 62:795–815.

Rzedowski, Jerzy, and Miguel Equihua

1987   *Atlas cultural de México: Flora.* Mexico City: Secretaría de Educación Pública, Instituto Nacional de Antropología e Historia, Grupo Editorial Planeta.

Sahagún, Bernardino

1982   *Historia general de las cosas de Nueva España.* 5th ed. Edited by Angel Maria Garibay. Mexico City: Editorial Porrua. Originally published 1793.

Samano Tajonar, Laura

n.d.   *Plantas y yerbas curativas de Mexico.* Mexico City: Gómez Gómez.

Sanecki, Kay N.

1992   *History of the English Herb Garden.* London: Ward Lock Villiers House.

Santamaría, Francisco J.

1978   *Diccionario de mejicanismos.* 3rd ed. Mexico City: Editorial Porrua.

Saunders, Lyle

1954   *Cultural Differences and Medical Care: The Case of the Spanish-Speaking People of the Southwest.* New York: Russell Sage Foundation.

Schauenberg, Paul, and Ferdinand Paris
  1977   *Guide to Medicinal Plants.* New Canaan, Conn.: Keats.
Schmidt, Alexander M.
  1990   Registration of Traditional Plant Remedies. In *Economic and Medicinal Plant Research.* Vol. 4, *Plants and Traditional Medicine,* ed. H. Wagner and N. R. Farnsworth. New York: Academic Press.
Schönfelder, Peter, and Ingrid Schönfelder
  1981   *Der Kosmos-Heilpflanzenfürer.* Stuttgart: Gesellschaft der Naturfreude, Franckh'sche Verlagshandlung.
Schuler, Irmgard
  1973   *Das "Florilegio Medicinal" von 1712 des Johann Steinhöfer in Mexiko.* Ph.D. diss., Ludwig-Maximilians-University, Munich, Germany.
Shutler, Mary Elizabeth
  1967   *Persistence and Change in the Health Beliefs and Practices of an Arizona Yaqui Community.* Ph.D. diss., University of Arizona.
  1977   Disease and Curing in a Yaqui Community. In *Ethnic Medicine in the Southwest,* ed. E. H. Spicer. Tucson: University of Arizona Press.
Segesser, Fillippe
  n.d.   Unedited letters, translated by Dan S. Matson.
Sessé, Martín [Martino] de, and Jose Mariano [Josepho Marianno] Moziño [Mociño, Mocinno]
  1887   *Plantae novae hispanae.* Mexico City: Oficina Tipografica de la Secretaria de Fomento. Originally published 1787.
Shemluck, Melvin
  1982   Medicinal and Other Uses of the Compositae by Indians in the United States and Canada. *Journal of Ethnopharmacology* 5:303–58.
Simpson, Lesley Byrd (editor and translator)
  1961   *Journal of Jose Longinos Martinez: Notes and Observations of the Naturalist of the Botanical Expedition in Old and New California and the South Coast, 1791–1792.* San Francisco: John Howell Books.
Singer, Merrill
  1989   The Coming of Age of Critical Medical Anthropology. *Social Science and Medicine* 28:1193–1203.
Siraisi, Nancy G.
  1985   *The Changing Fortunes of a Traditional Text: Goals and Strategies in Sixteenth Century Latin Editions of the Canoy of Avicenna.* Cambridge: Cambridge University Press.
  1990   *Medieval and Early Renaissance Medicine.* Chicago: University of Chicago Press.
Sloane, Hans
  1707   A Voyage to the Islands Madera, Barbados, Nieves, S. Christophers and JAMAICA with the Natural History . . . by Hans Sloane, MD, Fellow of the College of Physicians and Secretary of the Royal-Society. 2 vols. Sloane manuscripts 2941 and 3323. British Library.
Sobarzo, Horacio
  1966   *Vocabulario Sonorense.* Mexico City: Editorial Porrua.
Solecki, Ralph
  1975   Shanidar IV, a Neanderthal Flower Burial in Northern Iraq. *Science* 190: 880–81.

Somolinos d'Ardois, Germàn

1979   *Capitulos de historia medica mexicana.* Vol. 2, *El fenomeno de fusion cultural y su trascendencia medica,* and vol. 4, *Relacion y estudio de los impresos medicos mexicanos redactados y editados desde 1521 a 1618.* Mexico City: Dr. Juan Somolinos Palencia.

Spicer, Edward H.

1962   *Cycles of Conquest.* Tucson: University of Arizona Press.

1980   *The Yaquis.* Tucson: University of Arizona Press.

1983   Yaqui. In *Handbook of North American Indians,* vol. 10, vol. ed. A. Ortiz. Washington, D.C.: Smithsonian Institution Press.

Spicer, Edward H., and Raymond H. Thompson (editors)

1972   Introduction. In *Plural Society in the Southwest.* New York: Interbook.

Stearn, William T.

1990   *Botanical Latin.* New edition. London: David and Charles.

Steele, Arthur Robert

1964   *Flowers for the King.* Durham, N.C.: Duke University Press.

Steinbock, R. Ted

1985   The History, Epidemiology, and Paleopathology of Kidney and Bladder Stone Disease. In *Health and Disease in the Prehistoric Southwest,* ed. C. F. Merbs, R. J. Miller, and E. Alcauskas. Anthropological Research Papers No 34. Tempe: Arizona State University.

Stevenson, Matilda Coxe

1915   Ethnobotany of the Zuñi Indians. *Thirtieth Annual Report of the Bureau of American Ethnology,* pp. 31–64. Washington, D.C.: Government Printing Office.

Sub-Agencia de Agricultura y Ganaderia

1972   *Vegetación del estado de Sonora.* Hermosillo.

Tamayo, Joseph

1784   Relacion desde Arivechi, Sonora. WMS Amer. 49. Wellcome Institute for the History of Medicine, London.

Tewari, M. N.

1979   The Distribution of Medicinal Plants in the Arid and Semi-Arid Regions of Rajasthan-Thar Desert. In *Arid Land Plant Resources: Proceedings of the International Arid Lands Conference on Plant Resources,* ed. J. R. Goodin and D. K. Northington. Lubbock, Texas: International Center for Arid and Semi-Arid Land Studies.

Timbrook, Jan

1987   Virtuous Herbs: Plants in Chumash Medicine. *Journal of Ethnobiology* 7:171–80.

Timmerman, Barbara

1981   Larrea: Potential Uses. In *Larrea,* ed. E. Campos Lopez, T. J. Mabry, and S. Fernandez Tavizon. Mexico City: Consejo Nacional de Ciencia y Tecnología.

Tió, Aurelio

1966   *Dr. Diego Alvarez Chanca (Estudio Biográfico).* Instituto de Cultura Puertorriqueña, Universidad Interamericana de Puerto Rico. Barcelona: Imprime M. Pareja.

Train, Percy, J. R. Heinrichs, and W. Andrew Archer

1957   *Medicinal Uses of Plants by Indian Tribes of Nevada.* Facsimile of revised edition. Lawrence, Mass.: Quarterman Publications.

Treutlein, Theodore

1940   The Jesuit Missionary in the Role of Physician. *Mid-America* 22:120–41.

1945   The Relation of Filippe Segesser [1737]. *Iberoamerica* 27 (n.s. 16), no. 3:139–88.

1965   *Missionary in Sonora: The Travel Reports of Joseph Och, S.J.* San Francisco: California Historical Society.

Trigg, Heather B., Richard I. Ford, John G. Moore, and Louise D. Jessop

1994   Coprolite Evidence for Prehistoric Foodstuffs, Condiments, and Medicines. In *Eating on the Wild Side,* ed. N. L. Etkin. Tucson: University of Arizona Press.

Trotter, Robert T., II

1981a   Folk Remedies as Indicators of Common Illnesses: Examples from the United States–Mexico Border. *Journal of Ethnopharmacology* 4:207–21.

1981b   *Remedios Caseros:* Mexican American Home Remedies and Community Health Problems. *Social Science and Medicine* 15B:107–14.

Trotter, Robert T., II, and Juan Antonio Chavira

1981   *Curanderismo, Mexican American Folk Medicine.* Athens: University of Georgia Press.

Turner, William

1965   *Libellus de Re Herbaria* [1538] and *The Names of Herbes* [1548]. Facsimiles. London: Ray Society.

Tyler, Varro E.

1993   *The Honest Herbal.* 3rd ed. New York: Haworth Press.

Tyler, Varro E., Lynn R. Brady, and James E. Robbers

1981   *Pharmacognosy.* Philadelphia: Lea and Febiger.

Valdés, Javier, and Hilda Flores

1984   *Comentarios a la obra de Francisco Hernández.* Vol. 7 of *Obras completas.* Mexico City: Universidad Nacional Autónoma de México.

van der Eerden, M. Lucia

1948   *Maternity Care in a Spanish-American Community of New Mexico.* The Catholic University of America Anthropological Series no. 13. Washington, D.C.: Catholic University of America Press.

Velimirovic, Boris (editor)

1978   *Modern Medicine and Medical Anthropology in the United States–Mexico Border Population.* Pan American Health Organization Scientific Publication 359, Washington D.C.: Pan American Health Organization.

Veniaminov, Ivan [under Innokenti]

1984   *Notes on the Islands of the Unalashka District.* Translated by Lydia T. Black and R. H. Geoghegan, edited by Richard A. Pierce. Kingston, Ontario: Limestone Press. Originally published 1840.

Wagner, Charles John

1936   Medical Practices of the Yaquis. In *Studies of the Yaqui Indians of Sonora, Mexico,* ed. W. C. Holden, C. C. Seltzer, R. A. Studhalter, C. J. Wagner, and W. G. McMillan. Texas Technological College Bulletin vol. 12, no. 1. Lubbock.

Weber, Steven A., and P. David Seaman (editors)

1985   *Havasupai Habitat.* Tucson: University of Arizona Press.

Weil, Andrew

1990   *Natural Health, Natural Medicine.* Boston: Houghton Mifflin.

1995   *Spontaneous Healing.* New York: Alfred A. Knopf.

Weiss, Rudolf Fritz

    1988   *Herbal Medicine.* Translated by A. R. Meuss. Beaconsfield, U.K.: Beaconsfield
            Publishers.

Weller, S. C., L. M. Pachter, R. T. Trotter II, and R. D. Baer

    1993   Empacho in Four Latino Groups: A Study of Intra- and Inter-Cultural Vari-
            ation in Beliefs. *Medical Anthropology* 15:109–36.

Werner, David

    1970   Healing in the Sierra Madre. *Natural History* 79:60–70.

White, Stephen S.

    1948   The Vegetation and Flora of the Region of the Rio de Bavispe in North Cen-
            tral Sonora, Mexico. *Lloydia, a Quarterly Journal of Biological Science* 11, no.
            4:229–302.

Wilcox, David R.

    1986   The Tepiman Connection: A Model of Mesoamerican-Southwestern Inter-
            action. In *Ripples in the Chichimec Sea,* ed. F. J. Mathien and R. H. McGuire.
            Carbondale: Southern Illinois University Press.

Williams, Anita Alvarez de

    1983   Unpublished collection of Yaqui ethnobotany.

Winkelman, Michael

    1986   Frequently Used Medicinal Plants in Baja California Norte. *Journal of*
            Ethnopharmacology 18:109–31.

    1989   Ethnobotanical Treatments of Diabetes in Baja California Norte. *Medical*
            Anthropology 11:255–68.

Woosley, Anne I., and John C. Ravesloot

    1993   *Culture and Contact: Charles C. DiPeso's Gran Chichimeca.* Amerind Foun-
            dation Publications no. 2. Albuquerque: University of New Mexico Press.

Ximénez, Francisco

    1888   *Los quatro libros de la naturaleza y virtudes de las plantas.* Compiled by Nico-
            lás León. Morelia. Originally published 1615.

Ybarra, A. Fernandez de

    1894   The Medical History of Christopher Columbus. *Journal of the American*
            Medical Association 22 (May 5):647–54.

    1906   A Forgotten Worthy: Dr. Diego Alvarez Chanca, of Seville, Spain. *Journal of*
            the American Medical Association 47, no. 13 (September 29):1013–17.

Zamora-Martinez, Marisela C., and Cecilia Nieto de Pascual Pola

    1991   Medicinal Plants Used in Some Rural Populations of Oaxaca, Puebla, and
            Veracruz, Mexico. *Journal of Ethnopharmacology* 35:229–57.

Zigmond, Maurice L.

    1981   *Kawaiisu Ethnobotany.* Salt Lake City: University of Utah Press.

INDEX

Numbers in italics refer to illustrations

ABOUT THE AUTHOR

MARGARITA ARTSCHWAGER KAY is Professor Emerita at the University of Arizona. She received nursing degrees from Stanford University and the University of California, San Francisco, before practicing as a public health nurse in New York City and then serving as a teacher of maternal and child nursing in Tucson. She earned a Ph.D. in anthropology at the University of Arizona and fused both disciplines, as a medical anthropologist, in her teaching of courses such as Clinical Anthropology and Anthropology of Childbearing. Her research has focused on the health of Mexican American women through the life cycle. Her publications include "Health in a Mexican American Barrio," in *Ethnic Medicine in the Southwest,* ed. E. H. Spicer; *Southwestern Medical Dictionary* (Spanish-English, English-Spanish), now being expanded; and *Anthropology of Human Birth.* She is a Fellow of the American Anthropological Association and the American Academy of Nursing.

Plant drawings by Lucretia B. Hamilton. Reprinted, by permission, from Kittie F. Parker, *An Illustrated Guide to Arizona Weeds* (Tucson: The University of Arizona Press, 1972); Ervin M. Schmutz, Barry N. Freeman, and Raymond E. Reed, *Livestock Poisoning Plants of Arizona* (Tucson: The University of Arizona Press, 1985); and Lyman Benson and Robert A. Darrow, *Trees and Shrubs of the Southwestern Deserts* (Tucson: The University of Arizona Press, 1992). Figure 1 by Patricia B. Mason. Reprinted, by permission, from Charles T. Mason, Jr., and Patricia B. Mason, *A Handbook of Mexican Roadside Flora* (Tucson: University of Arizona Press, 1987).

Library of Congress Cataloging-in-Publication Data

Kay, Margarita Artschwager.
Healing with plants in the American and Mexican West / Margarita
Artschwager Kay ; with a foreword by Andrew Weil.
p.   cm.
Includes bibliographical references (p.) and index.
ISBN 0-8165-1645-6 (cloth : acid-free). — ISBN 0-8165-1646-4
(paper : acid-free)
1. Materia medica, Vegetable—West (U.S.).   2. Materia medica,
Vegetable—Mexico.   I. Title.
RS172.W4K38   1996
615'.32'0978—dc20   96-10101
                                                               CIP